Anaesthesia– Innovations in Management

Edited by
R. Droh, W. Erdmann, and R. Spintge

With 101 Figures and 22 Tables

Springer-Verlag
Berlin Heidelberg GmbH

Roland Droh
Sportkrankenhaus Hellersen
Lüdenscheid/Federal Republic of Germany

Wilhelm Erdmann
Erasmus Universiteit Rotterdam
Rotterdam/The Netherlands

Ralph Spintge
Sportkrankenhaus Hellersen
Lüdenscheid/Federal Republic of Germany

ISBN 978-3-540-13961-4   ISBN 978-3-642-82392-3 (eBook)
DOI 10.1007/978-3-642-82392-3

Library of Congress Cataloging in Publication Data. Main entry under title:
Anaesthesia: innovations in management. Includes index. 1. Anesthesia. I. Droh, R.
(Roland) II. Erdmann, W. (Wilhelm), 1940– . III. Spintge, R. [DNLM:
1. Anesthesia. WO 200 A529] RD81.A53 1985 617′.96 84-26740

Typesetting: Brühlsche Universitätsdruckerei, Giessen
Bookbinding: B. Helm, Berlin
2119/3020-543210

List of Contributors

W. A. van Alphen, Dep. of Plastic Surgery, Erasmus University,
3000 DR Rotterdam, The Netherlands

J. Baum, Anaesthesie Abteilung, Krhs. St. Elisabeth-Stift, Große Straße 54,
2845 Damme-Dümmer. Federal Republic of Germany

P. C. W. Beatty, Dep. of Anaesthesia, University Hospital of South Manchester,
Withington, Manchester, M20 8LR, United Kingdom

H. J. Beekman, Dep. of Anaesthesia, Acad. Ziekenhuis, Leiden, The Netherlands

H. J. Bender, Inst. f. Anaesthesiologie und Reanimation, Klinikum der Stadt
Mannheim, Theodor-Kutzer-Ufer, 6800 Mannheim, Federal Republic of
Germany

A. v. Bijnen, Dep. of Anaesthesia, Acad. Ziekenhuis, Leiden, The Netherlands

H. Boch-Fiola, Zentrum Anaesthesiologie der Universität Göttingen,
Robert-Koch-Str. 40, 3400 Göttingen, Federal Republic of Germany

K. Bonnhard, Anaesthesie Abteilung, St. Marien Krankenhaus,
Richard-Wagner-Straße 14, 6000 Frankfurt, Federal Republic of Germany

B. van der Borden, Dep. of Anaesthesia, Thorax Center, Erasmus University,
3000 Rotterdam, The Netherlands

H. Burchardi, Zentrum Anaesthesiologie der Universität Göttingen,
Robert-Koch-Str. 40, 3400 Göttingen, Federal Republic of Germany

A. G. L. Burm, Dep. of Anaesthesia, Acad. Ziekenhuis, Leiden, The Netherlands

M. D'Ambra, Dep. of Anaesthesia, Harvard Medical School, Boston, MA USA

M. Dhont, Dep. of Anaesthesia, Acad. Hospital, 9000 Ghent, Belgium

R. Droh, Anaesthesiologie, Sportkrankenhaus Hellersen, 5880 Lüdenscheid,
Federal Republic of Germany

R. Dudziak, ZAW, Klinikum d. J. W. Goethe Universität, 6000 Frankfurt,
Federal Republic of Germany

W. L. den Dunnen, Dep. of Anaesthesia, Municipal Hospital Dordrecht,
Dordrecht, The Netherlands

B. Dworacek, Dep. of Anaesthesia, Erasmus University, 3000 DR Rotterdam, The Netherlands

W. Erdmann, Dep. of Anaesthesia, Erasmus University, 3000 DR Rotterdam, The Netherlands

C.E. Essed, Dep. of Pathology I, Erasmus University, 3000 DR Rotterdam, The Netherlands

N.S. Faithfull, Dep. of Anaesthesia, Erasmus University, 3000 DR Rotterdam, The Netherlands

M. Fenema, Dep. of Anaesthesia, Erasmus University, 3000 DR Rotterdam, The Netherlands

H. Förster, ZAW, Klinikum d. J. W. Goethe Universität, 6000 Frankfurt, Federal Republic of Germany

P.A. Foster, Dep. of Anaesthesia, University of Stellenbosch, Tygerberg, South Africa

E. Freis, Dep. of Anaesthesia, Harvard Medical School, Boston, MA USA

S. Gonserowski-Spintge, Im Brantenberg 1, 5883 Kierspe, Federal Republic of Germany

D. de Haas, Dep. of Anaesthesia, Acad. Ziekenhuis, Leiden, The Netherlands

I. Hänsel, Zentrum Anaesthesiologie der Universität Göttingen, Robert-Koch-Str. 40, 3400 Göttingen, Federal Republic of Germany

T. von dem Hagen, Dräger-Werk AG, Moislinger Allee 53, 2400 Lübeck, Federal Republic of Germany

H.J. Hartung, Inst. f. Anaesthesiologie und Reanimation, Klinikum der Stadt Mannheim, Theodor-Kutzer-Ufer, 6800 Mannheim, Federal Republic of Germany

H.-D. Hattendorff, Dräger-Werk AG, Moislinger Allee 53, 2400 Lübeck, Federal Republic of Germany

T.E.J. Healy, Dep. of Anaesthesia, University Hospital of South Manchester, Withington, Manchester, M20 8LR, United Kingdom

H.-P. Kamin, Corvinusgasse 4, 1038 Vienna, Austria

B. Kay, Dep. of Anaesthesia, University Hospital of South Manchester, Withington, Manchester, M20 8LR, United Kingdom

G. Klein, ZAW, Klinikum d. J. W. Goethe Universität, 6000 Frankfurt, Federal Republic of Germany

S. Klein, Anaesthesie Abteilung, St. Marien Krankenhaus, Richard-Wagner-Straße 14, 6000 Frankfurt, Federal Republic of Germany

L. Kleinschmidt, Dräger-Werk AG, Moislinger Allee 53, 2400 Lübeck, Federal Republic of Germany

P. Knegt, Dep. of Ears-Noso-Throat Surgery, Erasmus University, Acad. Hospital Rotterdam, 3000 DR Rotterdam, The Netherlands

E. Lachitjaran, Dep. of Anaesthesia, Erasmus University, 3000 DR Rotterdam, The Netherlands

R. Lapin, Institute of Bloodless Medicine and Surgery, 707 E. Chapman Ave, Fullerton, CA 92631, USA

H. Lutz, Inst. f. Anaesthesiologie und Reanimation, Klinikum der Stadt Mannheim, Theodor-Kutzer-Ufer, 6800 Mannheim, Federal Republic of Germany

V. Machaj, Dep. of Anaesthesia, Harvard Medical School, Boston, MA USA

C. Mallios, Dep. of Anaesthesia, Erasmus University, Acad. Hospital Rotterdam, 3000 DR Rotterdam, The Netherlands

J. W. R. McIntyre, Dep. of Anaesthesia, University Hospital, Edmonton, Alberta, Canada

S. Meij, Dep. of Anaesthesia, Thorax Center, Erasmus University, 3000 DR Rotterdam, The Netherlands

T. Mostert, Dep. of Anaesthesia, Municipal Hospital Dordrecht, Dordrecht, The Netherlands

G. Müller, Umweltbundesamt, Bismarckplatz 1, 1000 Berlin 33, Federal Republic of Germany

P. M. Osswald, Inst. f. Anaesthesiologie und Reanimation, Klinikum der Stadt Mannheim, Theodor-Kutzer-Ufer, 6800 Mannheim, Federal Republic of Germany

U. Ottermann, Anaesthesie Abteilung, St. Marien Krankenhaus, Richard-Wagner-Straße 14, 6000 Frankfurt, Federal Republic of Germany

T. Oyama, Dep. of Anaesthesia, Hirosaki University School of Medicine, Hirosaki, Japan

D. M. Philbin, Dep. of Anaesthesia, Harvard Medical School, Boston, MA USA

O. Prakash, Dep. of Anaesthesia, Thorax Center, Erasmus University, 3000 Rotterdam, The Netherlands

F. de Raadt, Dep. of Anaesthesia, Acad. Ziekenhuis, Leiden, The Netherlands

W. Rating, Dep. of Anaesthesia, Erasmus University, 3000 DR Rotterdam, The Netherlands

V. Rejger, Dep. of Anaesthesia, Acad. Ziekenhuis, Leiden, The Netherlands

G. Rolly, Dep. of Anaesthesia, Acad. Hospital, 9000 Ghent, Belgium

C. E. Rosow, Dep. of Anaesthesia, Harvard Medical School, Boston, MA USA

M. R. van Santen, Dep. of Gynaecology, Utrecht University Hospital, Utrecht, The Netherlands

P. A. Scheck, Dep. of Anaesthesia, Erasmus University, Acad. Hospital Rotterdam, 3000 DR Rotterdam, The Netherlands

R. M. Schepp, Dep. of Anaesthesia, Erasmus University, 3000 DR Rotterdam, The Netherlands

R. C. Schneider, Dep. of Anaesthesia, Harvard Medical School, Boston, MA, USA

U. Schneider, Anaesthesie Abteilung, Krhs. St. Elisabeth-Stift, Große Straße 54, 2845 Damme-Dümmer, Federal Republic of Germany

R. Serreyn, Dep. of Anaesthesia, Acad. Hospital, 9000 Ghent, Belgium

C. E. van der Smissen, Dräger-Werk AG, Moislinger Allee 53, 2400 Lübeck, Federal Republic of Germany

A. R. Smith, Dep. of Plastic Surgery, Erasmus University, 3000 DR Rotterdam, The Netherlands

J. Spierdijk, Dep. of Anaesthesia, Acad. Ziekenhuis, Leiden, The Netherlands

R. Spintge, Anaesthesiologie, Sportkrankenhaus Hellersen, 5880 Lüdenscheid, Federal Republic of Germany

L. M. Stanford, Dep. of Anaesthesia, University Hospital, Edmonton, Alberta, Canada

T. Stokke, Zentrum Anaesthesiologie der Universität Göttingen, Robert-Koch-Str. 40, 3400 Göttingen, Federal Republic of Germany

A. Trouwborst, Dep. of Anaesthesia, Erasmus University, 3000 DR Rotterdam, The Netherlands

L. Versichelen, Dep. of Anaesthesia, Acad. Hospital, 9000 Ghent, Belgium

C. F. Wallroth, Dräger-Werk AG, Moislinger Allee 53, 2400 Lübeck, Federal Republic of Germany

L. Weller, Inst. f. Anaesthesiologie und Reanimation, Klinikum der Stadt Mannheim, Theodor-Kutzer-Ufer, 6800 Mannheim, Federal Republic of Germany

B. Westerkamp, Mijnhardt b. v., Singel 45, 3984 NV Odijk, The Netherlands

H. Yanagida, Dep. of Pain Clinic Anaesthesia, The Tokyo Nenkin Hospital, Tokyo University, 162 Tokyo, Japan

H. T. van der Zee, Dep. of Anaesthesia, Albany Medical College, Albany, NY 12208, USA

Contents

X

Introduction

R. Droh

Ladies and gentlemen, dear friends and colleagues, we welcome you very cordially to our symposium "Innovations in Management and Technic and Pharmacology". We are very glad that you have come to Lüdenscheid and we do hope that our programme will fulfil your expectations.

We decided to hold this symposium, because it is getting more and more difficult to select innovations at international congresses around the world which are important for our clinical work. Now and in the future our intention is to present the actual state of technology, management and pharmacology. We would be very glad to receive your suggestions for further symposia.

The industry has the same problems as we have. They do not only have to search for those things which can be realized and which are desirable, but also for those things which can be sold. But the industry must also be stimulated by the inventors and by the users, so we want to bring together the industry, the physicians and the inventors for fruitful discussions. And we hope that in the future the industry will provide us more quickly with those technical and organizational aids that we need.

We want the industry no longer running behind the market but heading the market. At present too many interesting developments are killed by so-called market analysis, in the beliefing that such analysis can always prove what cannot be sold. In anaesthesia many companies are always busy with the same product. Decisive new ideas are neglected as soon as they seem not to be "in". The only exception is the pharmaceutical industry.

We therefore want to encourage companies to try out fundamental new ideas and we want to keep them away from their anxious and purely mercantile thinking about new developments. This wish is especially directed towards the big potent and leading companies.

At the same time we intend to bring together all those who do have creative thoughts and who are looking for some help in promoting them. In my opinion this includes bringing together the single specialties of theoretical and clinical medicine to discuss their newest developments and to make them understandable. In this way the innovations can be brought more rapidly into industrial production and into use. So it might also be possible to write less. I think we already have to read too many special texts which are basically not "readable" and too often even not understandable because they are too specialized.

Nowadays, rationalization, considering the decisive parameters of our dialy clinical tasks, is a basic necessity for straight medical work. Another fundamental

1

requirement is a skilled handling of the instruments. Of course, we also always have to consider the relationship of costs and practical relevance. If we ware freed from huge amounts of useless beaurocracy, we will have more time to think over important problems, to read and to discuss. As we are also managers today, innovations can support and relieve us.

The Closed-Circuit Anaesthesia System

R. Droh

The closed–circuit system is certainly the most modern anaesthesia system presently in use. We have already been using it for a period of 10½ years without any problems and in this time we have dealt with more than 50 000 patients.

The superiority of this system can be summarized in 16 points:

1. The narcotic gases are warm.
2. The narcotic gases are perfectly humidified.
3. The oxygen consumption per minute and per breath can each be measured exactly.
4. Not only can the CO_2 deriving from the oxygen consumption be calculated quantitatively per minute and per breath, but it can also be measured exactly.
5. Changes in lung perfusion can be recognized immediately.
6. Changes in cardiac output can be recognized immediately. It will be possible to calculate these changes via non-invasive measurements in the near future.
7. The ventilation of the patient can be controlled exactly, i. e. true normoventilation can be achieved for every breath. Hypo- and hyperventilation are thus excluded. As far as hyperventilation is necessary for certain operations, it can be applied exactly.
8. Changes in blood pH can be measured and can be calculated approximately.
9. At present the depth of narcosis in inhalation anaesthesia can be estimated most exactly with the closed-circuit system.
10. The uptake of volatile anaesthetics such as halothane, isoflurane and methoxyflurane can be measured exactly.
11. The concentration of O_2 and N_2O can approach the normal O_2 concentration in the air because O_2 consumption as well as CO_2 output can be controlled exactly and can be kept in the normal range. For more than 1 year we have been anaesthetizing with about 24 vol.% O_2 and 76 vol.% N_2O inspiratory and about 20 vol.% O_2 and 80 vol.% N_2O end-expiratory.
12. The elimination of excess narcotic gases is unnecessary.
13. It is possible to recognize pathological changes in metabolism at once. One does not have to rely upon secondary clinical signs.
14. The environmental pollution through the permanent outflow of excess gases into the atmosphere via leaks in the system is not only nearly zero in the operations-theatres but it is also drastically reduced in the outer atmosphere.

15. These advantages do not involve any costs; the system even saves additional money. It is not only much better and much more secure than all the others, but at the same time it is more economic.
16. It is the only anaesthetic system which can be fully computer automatized without any risk to the patient.

Points 1,2. The anaesthetic gases are warm and humidified. Here we should remember that after a period of only 5 min with a humidification of only 70 vol.% H_2O the action of the cilia of the respiratory tract slows down. With 50 vol.% H_2O the cilia no longer work after a period of 8–10 min. And with 30 vol.% H_2O they stand still after only 3–5 min.

Everybody must realize how uncomfortable he feels when in winter-time he is forced to inhale cold and dry air. Many people soon catch cold. Just after a few breaths the nose hurts, so that one begins to breathe through the open mouth. Often one covers one's mouth and nose with one's hands in order to warm up the cold air.

The same is true for dry air. In a desert one feels pain if one is forced to inhale warm but dry air. This is the reason why people living in these regions cover their mouth and nose with a shawl, to humidify the air by the water kept back in the shawl. The combination of coldness and dryness in the air intensifies the negative effect of both.

In the semi-closed anaesthesia system the patient receives a minimum of 1000 ml O_2 and 2000 ml N_2O per minute, mostly 2000 ml O_2 and 4000 ml N_2O per minute. In the open systems even 4000 ml O_2 and 8000 ml N_2O per minute are administered. These gases are cold and dry and the poor patient has no chance to reject these gases.

Anaesthesia itself depresses the whole metabolism. A reduced metabolism is accompanied by a reduced immunological defence. Furthermore, the cilia-carrying epithelium is depressed by the dry and cold gases and dried out by our atropine medication. Therefore, the microbes have a good opportunity to invade the lung tissue. We do not only treat adults in this way, but also children and neonates. Out of the huge amount of gases they only can use 10–14 ml O_2 per minute. Moreover, the newborn are most sensitive to the loss of warmth and humidity.

With the closed-circuit system we can avoid all these hazards. And no equipment is needed to warm up or to humidify the gases.

Incidentally, it seems to me important to consider that even with the available equipment it is not possible to humidify and warm up the high gas flows of the semi-open or semi-closed systems sufficiently!

How much warmth the body can lose when being ventilated with cold gases is demonstrated by the technique of hypothermia. In my opinion this is one reason for the postoperative shivering of adults and the rapid drop in body temperature, especially in children and neonates.

Our system not only provides the patients with warm and humid gases, but it is also a very simple system, and should also be used for automatic ventilation in intensive care patients.

Since we have been using closed-circuit systems we have not seen any postoperative lung complications. This means that no postoperative inhalation therapy is necessary. I am quite sure that many lung complications involved in prolonged ventilation would not occur if closed-circuit systems were always used instead of semi-open or semi-closed systems.

Points 3,4,5,6,13. With this system oxygen consumption and CO_2 production can be measured exactly breath by breath and changes in lung perfusion and changes in cardiac output can be recognized immediately. No other system enables oxygen consumption to be measured so easily because huge amounts of oxygen are administered to the patient, though only small percentages of it are taken up. In view of the present states of technology this is an anachronistic waste of money and resources.

From the oxygen consumption one can get a lot of very interesting information about the patient, for example by Fick's formula:

$$CMV = \frac{\text{oxygen consumption (ml/min)} \times 100}{avDO_2 \text{ (vol. \%)}},$$

CMV = cardiac output per minute; $avDO_2$ = arterio venous O_2 difference. If the CMV is divided by the pulserate, this gives the stroke volume of the heart beat by beat. I am sure that it will soon become possible to calculate the $avDO_2$ by means of biomathematical models and without catheters – Probably with a wash-out method employing N_2O and/or halothane. This would represent direct information about the efficiency of the heart, far superior to all parameters measurable at the moment. So the oxygen consumption gives us basic information about metabolism and changes of metabolism. Furthermore, O_2 uptake and CO_2 output convey information about changes in lung perfusion.

Decrease in cardiac output leads to a decrease in oxygen consumption and oxygen uptake into the tissues. So after cardiac arrest it can be seen that the patient no longer takes up oxygen and does not produce any CO_2. It is very simple to recognize insufficient heart massage and to improve it with this system. A reduction of the necessary O_2 uptake of the patient is directly proportional to the reduction of the cardiac output, which is congruent with a parallel measuring of the CO_2 production.

Pathological changes in metabolism can be recognized at once: it is no longer necessary to wait for secondary clinical signs. The start of hyperthermia, for instance, can be seen at once by the increasing oxygen uptake combined with an increase of the CO_2 production. For there is no big fire without equivalent of oxygen. Also in hyperthyreosis more oxygen is taken up and CO_2 produced in relation to the body weight, but not as much as in the case of hyperthermia.

Of course, one can also obtain a precise view into the normal metabolism by means of the respiratory quotient (RQ). When we have pure oxidation of carbohydrates the consumed oxygen is transformed into the same amount of CO_2. So

$$RQ = V_{CO_2}/V_{O_2} = 1.$$

If there is a pure utilization of fat, then O_2 uptake is higher than CO_2 output because oxygen is not only necessary for the oxidation of carbon but also for the oxidation of hydrogen. RQ therefore is beyond 0.7. Pure utilization of proteins has a RQ of 0.8.

Changes in the RQ also occur as a result of hyperventilation or increases in muscle activity (e.g. fasciculation after succinyl dicholine). None of these variationes can be seen with the usual coarse surplus systems.

Point 7. The ventilation of the patient is much more exactly steerable than with the semi-closed or open techniques, i.e. we can have a true normoventilation. If hyperventilation is desired, it is possible to apply it in an exact and quantitative way.

One great danger I see in the usual anaesthesia systems is hyperventilation. Some 20–30 years ago our main problem was to avoid hypoventilation. Nowadays hyperventilation is the major risk. Erdmann and I call this hyperventilation the "neuron killer" of today.

The importance of normoventilation cannot be discussed here in detail. I only want to draw attention to the Bohr effect, with its importance for the whole O_2-CO_2 exchange and the acid-base-H_2-O- electrolyte system and its influence on the buffer capacity.

On the whole I do not think the importance of hyperventilation for the cerebral blood flow, the intra-operative and postoperative steering of respiration, and the changes in intracranial pressure is fully realized.

Another advantage of the closed-circuit system with its normoventilation is the almost complete elimination of postanaesthetic vomiting. In my opinion, the effect of normoventilation during the whole course of anaesthesia is to prevent a reactive postoperative hypoventilation. As this system also enables us to control the depth of anaesthesia more exactly, the postoperative respiratory depression is less.

Point 8. Changes in blood pH, for instance after the application of a tourniquet, result from the reduced oxygen requirement in the ischaemic tissue and the increase in CO_2 after the reperfusion of this tissue. These changes can be measured in the closed-circuit system, so ventilation can be controlled according to the O_2 and CO_2 values in order to achieve normal pH levels in the blood by increasing CO_2 output. This way buffer capacity is released and accompanying changes in electrolyte levels are balanced.

Point 9. With the usual anaesthesia systems the depth of anaesthesia can only be estimated by means of pulse rate (plethysmography) and blood pressure. The only exception is ether. Also the EEG is no reliable parameter.

In the closed system the quantitative measurement of O_2 uptake and CO_2 output gives a certain indication of the depth of narcosis. Of course, one cannot differentiate between a deep and a very deep narcosis. But when the patient comes back through the state of tolerance then oxygen consumption increases. If the system is absolutely tight and if the calculated oxygen demand is surpassed, then this is a sign that anaesthesia is becoming lighter. This early information is a great help in controlling the depth of narcosis.

Point 10. As the system is closed, neither O_2 and N_2O nor any of the volatile anaesthetics are able to leave the system. The volatile anaesthetic circulates in the system until it is absorbed by the patient. So the decrease in the volume of the volatile anaesthetic in the vaporizer is directly related to the uptake of this anaesthetic by the patient. From the difference between output of the anaesthetic from the vaporizer and the end-expiratory concentration it is possible to tell exactly how much volatile anaesthetic has been absorbed per breath by the patient.

In this connection I would like to ask Dräger to improve theier vaporizer by adding a monitor indicating the volume of anaesthetic in the vaporizer. It should have an accuracy of about 0.1 ml.

Point 11. Besides a secure oxygen supply, the concentration of N_2O also has a certain importance. All animal life on earth has been adjusted to an oxygen concentration of about 21 vol.% for millions of years. For a long time anaesthaesiologists were anxious to provide patients with enough oxygen, and used not only a very high gas flow but also a high oxygen concentration in order to secure sufficient oxygen over supply. The usual mixture contains between 30 vol.% and 40 vol.% oxygen.

In the closed-circuit system it is now possible to control the uptake of oxygen and the exact oxygen concentration in the narcotic gases. So the oxygen concentration in the system can approach the oxygen concentration in the outer atmosphere. This way the N_2O concentration can be increased. For more than 1 year we have been using oxygen concentrations of about 24 vol.% and N_2O concentrations of about 76 vol.% inspiratorilly and about 20 vol.% oxygen and 80 vol.% N_2O expiratorilly. The effect of N_2O is markedly improved and the end-expiratory concentration of volatile anaesthetics can be reduced. The same is true for the necessary dosage of fentanyl in neuroleptanaesthesia.

For us it is no question of producing a pure oxygen narcosis with a volatile anaesthetic without using any N_2O. On the contrary: by getting near to 21 vol.% oxygen – similar to the outer atmosphere – with the concentration of O_2 in the narcotic gas mixture we increase the N_2O concentration. We can thus substantially improve the course of gas anaesthesias. We have been using this technique more than 1 year with great success. Using this method we need smaller amounts and lower concentrations of volatile anaesthetics. We can better control the recovery from anaesthesia. The patients awake very quickly.

Of course, an obese patient with a high body weight who is also a smoker and drinks regularly needs higher amounts of anaesthetics than a normal patient.

Points 12,14. I should like to draw attention to environmental pollution through the high and costly waste of gas with the open, semi-open and semi-closed systems. Basically, the patient needs only 3.5 ml oxygen per minute per kilogram and about 1.5 ml N_2O per minute per kilogram body weight. So a patient with a body weight of 70 kg needs about 250 ml oxygen and about 100 ml N_2O per minute at the beginning of anaesthesia. At the beginning the patient only demands about 350 ml of the gas mixture per minute. With the duration of anaesthesia the patient takes up less and less N_2O. The tissue gets more and more saturated with anaesthetics and so metabolism is depressed and less oxygen is needed.

Using 2 l oxygen and 4 l N_2O per minute in the semi-closed system only 6 vol.% of this amount is used by the patient. In the open system with a gas flow of 4 l O_2 and 8 l N_2O per minute, the excess in adults is 97 vol.% and in neonates 99.9 vol.%. This not only causes environmental pollution but is also a waste of resources!

Many people believe they can avoid these problems by installing pipelines to exhaust excess gases of the operating-theatres. But most of the excess gases remain in the operating-theatre because of the leaks in the system and in the ventilators: the pipelines are only able to exhaust a small amount of the excess gases. Those excess gases being exhausted from the operating-theatres go up into the higher atmosphere, or they come back with rain and fog.

Point 15. The waste of resources is not the only problem. Huge amounts of money are being wasted, as you can see from Figs. 1 and 2.

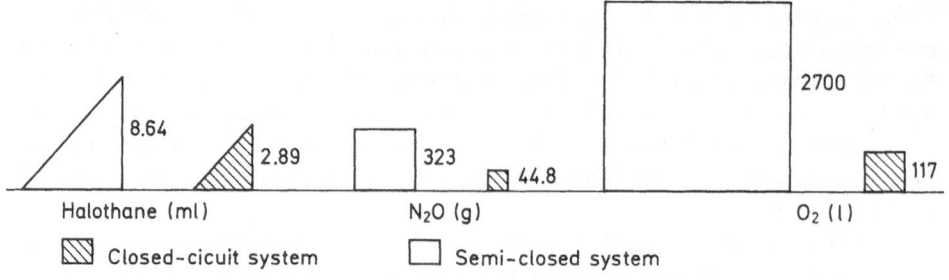

Halothane (ml)   N₂O (g)   O₂ (l)

[NN] Closed-cicuit system   [ ] Semi-closed system

Fig. 1. Use of gases by each patient under the closed-circuit and semi-closed systems

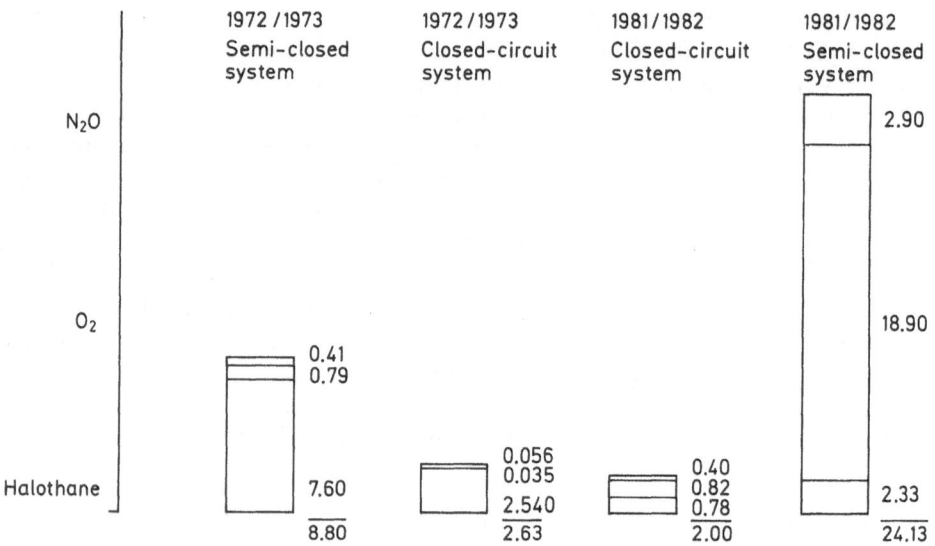

Fig. 2. Cost per patient of anaesthesia under the closed-circuit and semi-closed systems

Closed-circuit anaesthesia provides completely new insights into the control of anaesthesia, breath by breath, heart beat by heart beat. I am sure that there are a lot of advantages which we do not yet perceive. Let me therefore compare the open and the semi-open systems to an aeroplane from the very early days of flying; let me compare the semi-closed system to a plane from the 1930s or 1940s and let me compare the closed-circuit system to a modern, highly sophisticated aircraft, the development of which has just begun and will go on for a long period.

If anyone says that closed-circuit anaesthesia has been here before, then I would like to answer that before the two brothers Wright others tried to fly. They did fly but they did not have a happy landing: they fell down. Closed-circuit anaesthesia was dangerous until about 11 years ago. But now we have been able to "fly" with this system for more than 10½ years, with more than 50 000 "passengers" of all ages without any accident. We also fly not only when the weather is fine but also when it is severe. Our patients have all the problems with which modern anaesthesia has to deal.

References

Droh R (1976a) Das geschlossene Narkose-Kreissystem. Anaesth Praxis 11:27–32
Droh R (1976b) Clinical significance of recent research results. VI. World congress of anaesthesiology, Mexico, 24–30 April 1976, Programa oficial p 84, p 369 Convention Reporter, Anesthesiology edition, McNeil Laboratories, Fort Washington, PA 19034, vol 1 no 2
Droh R (1981) The closed circuit system and its advantages. A review of the last nine years. International symposium on closed circuit anaesthesia. 10–12 April 1981, Birmingham
Droh R (1982) Die Inhalationsnarkose im geschlossenen System. Internationales Symposion: geschlossenes System für Inhalationsnarkosen. 7–8 May 1982, Düsseldorf
Droh R, Erdmann W (1982) Closed circuit anaesthesia and its significance. VI. European Congress of Anaesthesiology. 8–15 September 1982, London
Droh R, Rothmann G (1977) Das geschlossene Kreissystem. Eine diskreditierte Konzeption der Vergangenheit, durch Innovationen die Methode der Wahl. Anaesthesist 26:461–466
Droh R, Spintge R, Erdmann W (1982) Die Überlegenheit des geschlossenen Kreissystems. Ein Zehnjahresbericht über 50000 Fälle. Deutscher Anaesthesie Congress 1982

Principles of Low Gas Flow Measurement for Closed-Circuit Systems

T. von dem Hagen and L. Kleinschmidt

In recent years an increased trend towards the completely closed-circuit system can be observed in anaesthesia technology. One of the important requirements for this development is an improvement in the technique of gas measurement so that the important parameters can be measured with adequate precision and sensitivity. One of these important parameters is the amount of fresh gas being metered.

One characteristic of the closed-circuit system is that only that amount of gas is metered that the patient actually takes up, in other words, that he requires. For metering these relatively small amounts of gas and for safety reasons, measurement and display of the flow rate is usually necessary as well.

In the area of flow measurement there are countless systems which utilize a variety of different physical effects. Therefore, in the following, only those methods will be mentioned which are suitable for measuring small amounts of gas flow. The most important methods and their principles of function are shown schematically in Fig. 1. There are several other promising systems which, however, have not yet been sufficiently developed.

In order then to find the optimum method of measurement for a given problem, it is first necessary to specify the requirements. In doing so it must be taken into account that the parameters of the state of a gas, such as pressure, volume and temperature, are strongly dependent on each other, as is generally known. Table 1 lists the specifications for a flow meter for metering fresh gas in a closed system.

Reasons for the individual items can be given as follows. The required amount of fresh gas, nitrous oxide or oxygen in a closed system is mainly determined by the body weight of the patient. Range I, as given, is sufficient in any case. In certain phases of the anaesthesia, however, it may also be necessary to open the system, which could mean that the precision might be lower than in range I. In this case, range II applies. The listed tolerance of $\pm 10\%$ is considered a minimum requirement and is part of the Draft International Standards Organization Standard. Within range I a lower value is certainly desireable.

More important than the absolute precision is, in many cases, the reproducibility and, connected with it, the stability of the set gas flow or its displayed amount. Such high demands cannot be made on the time resolution. It is necessary to make sure, however, that when a change is made in the gas flow, for instance by means of a manual metering valve, the display is not too greatly delayed.

In contrast to measurement in the breathing system, the pressure drop at the sensor in the fresh gas tube can be relatively large. It must be considered, however,

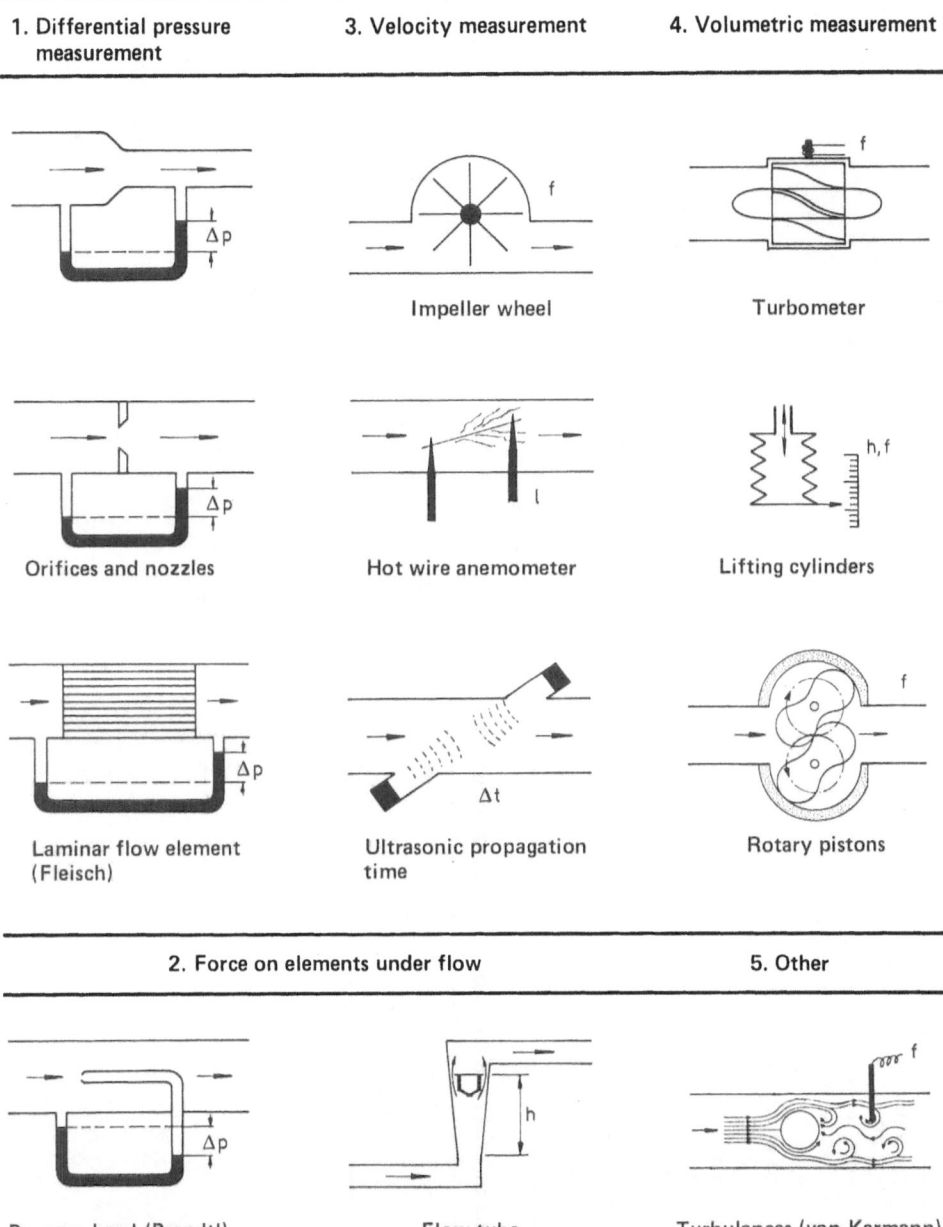

1. Differential pressure measurement	3. Velocity measurement	4. Volumetric measurement

Δp

Impeller wheel

Turbometer

Orifices and nozzles

Hot wire anemometer

Lifting cylinders

Laminar flow element (Fleisch)

Ultrasonic propagation time

Rotary pistons

| 2. Force on elements under flow | | 5. Other |

Pressure head (Prandtl)

Flow tube

Turbulances (von Karmann)

Fig. 1. Several principles of gas flow measurement

that due to compression before the sensor, an additional error may result. But such an error, at the given maximum value of 5%, is still sufficiently small.

The other items need no further explanation.

To obtain an overall evaluation of the individual methods relative to their specifications, data were compiled from the literature and from company brochures and a division into the three ratings (+, 0 and −) was made. Table 2 shows schematically, using these symbols, how the methods of potential interest behave.

Table 1. Specifications for a flow measurement device for metering fresh gas in a closed system

Range I	20–500 ml/min ⎫ dynamic range 1:500
Range II	0.5–10 l/min ⎭
Tolerance	±10% of the measured value
Reproducibility	±2% of the measured value
Time resolution	0.5 s
Flow resistance	50 mbar/10 l per minute
Temperature range	15°–40 °C
Electronic output signal	Yes
Fail-safe	Yes
Effects of gas type, moisture, dirt	None (slight)
Dependence on position	Slight
Hygiene, care, servicing	Slight
Technical resources, price	Slight

Table 2. Evaluation of several characteristics of commonly used flow-measuring methods

	Low Flow	Dynamic range		Accuracy	Repro-ducibility	Time resolution	Remarks
		1:500	1:25				
Differential pressure	+	–	–	+	+	+	
Fleisch's nozzle	+	–	–	+	+	+	Adequate for use only for laboratory analysis
Flow tube	+	–	0	+	0	+	Dependent on position
Impeller wheel	0	–	0	+	+	0	
Hot wire	+	0	+	+	+	+	
Ultrasonic	–	–	–	+	+	+	Technically elaborate
Turbo	0	–	0	+	+	0	
Lifting cylinders	+	0	+	+	+	0	Discontinuous
Rotary pistons	0	–	–	+	+	0	
Turbulances	0	–	–	+	+	+	Sensitive sensors

+ can be achieved; 0 can be achieved with additional technical resources; − cannot be achieved

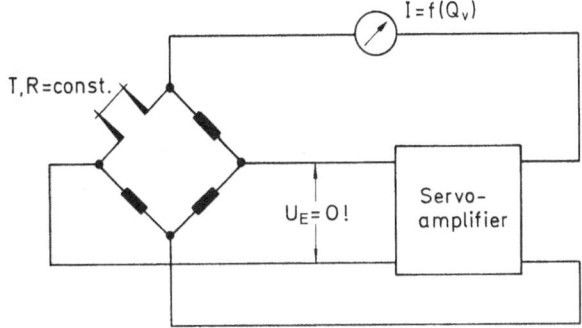

$I = f(Q_v)$

T, R = const.

$U_E = 0!$

Servo-amplifier

Fig. 2. Basic circuit diagram for a constant temperature anemometer. T, temperature; R, resistance; I, current; Q_v, flow; U_E, input voltage

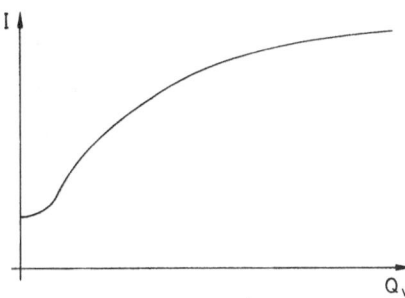

I

Q_v

Fig. 3. Basic calibration curve for a hot wire anemometer

The additional technological resources listed for "0" can imply, for example, a temperature or pressure compensation as well as more elaborate electronic or mechanical measures which, at a rating of "–", would not be economically sound.

One notices right away that the required dynamic range of 1:500 cannot be achieved by any of the methods without additional resources. Therefore, another column with a range of 1:25 was added which allows an evaluation in case the method in question should be interpreted for two overlapping ranges.

Thus, with the aid of Table 2, the number of potential measuring methods can be reduced to five. Only those methods which do not show a "–" in any column (except in column 2) appear in the discussion. These are the hot wire anemometer, lifting cylinders method, impeller wheel, turbometer and flow tubes.

These remaining methods will now bei discussed in detail. The basic circuit diagram for a hot wire anemometer is shown in Fig. 2. The constant temperature method is preferable. The servo-amplifier always sets its output voltage so that the input voltage is exactly $U_E = 0$. The output current is then a measure for the flow velocity of the gas.

The emission of heat results not only from the surrounding gas flow but also from radiation, heat conduction via the connecting pins and its own convection. This results in a complex relationship between the sensor current and the flow velocity in the gas. This causes problems, in particular with very low flows, and requires a calibration of the device. An electronic linearization of the indicator is a good idea. In Fig. 3 an example of a calibration curve is shown.

The hot wire anemometer offers a very good time resolution combined with high measuring sensitivity and a wide dynamic range of approximately 1:200.

The principle of the lifting cylinders meter is shown in greater detail in Fig. 4. The valve control by means of the piston position at the lower and upper turning

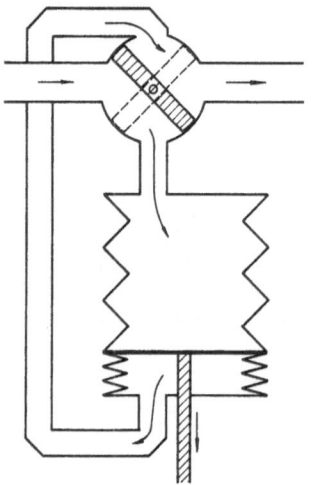

Fig. 4. Lifting cylinders flow meter

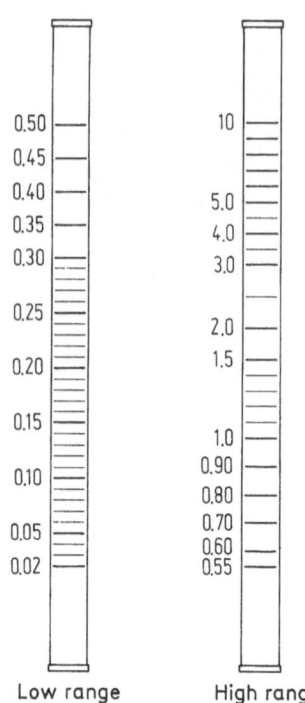

Low range   High range

Fig. 5. Scaling suggestion for a flow tube unit ▶

points is not shown. Newer devices work with several cylinders for smoother operation. Errors in measurement occur because of leakage and dead space during valve shifting.

By reducing the piston stroke and the bellows volume and by using longer measuring times, extremely low flows down to far below 1 ml/min can be measured very exactly. A dynamic range of 1:100 can be achieved without resorting to greater technical resources.

Impeller wheel meters and turbometers have very similar characteristics, since both work with rotating elements. At the start, forces of friction must be overcome in order to create a minimum velocity of flow for operation. This leads to problems in the measurement of very low flows, especially since the dynamic range is not very wide. Another disadvantage is their behaviour at sudden changes in the flow (delayed starting and slowing down respectively). But otherwise they are very reliable and precise.

Flow meters using floating elements are very widely used in medical technology, since they are very safe and reliable due to their simple, purely mechanical principle of function. The measurement cones are generally made of glass and are manufactured by special processes at very high precision. The shape of the floating elements has a very significant influence on the viscosity sensitivity of the flow meter.

Flow meters using floating elements have wider measurement tolerances and are more dependant on viscosity than the other methods mentioned. Another disadvantage is that the electronic display of the flow rate requires considerable additional resources. To achieve the required dynamic range of 1:500 using two tubes, special manufacturing requirements must be met. Fig. 5 shows a scaling suggestion for a flow meter unit in the fresh gas tube of a closed system.

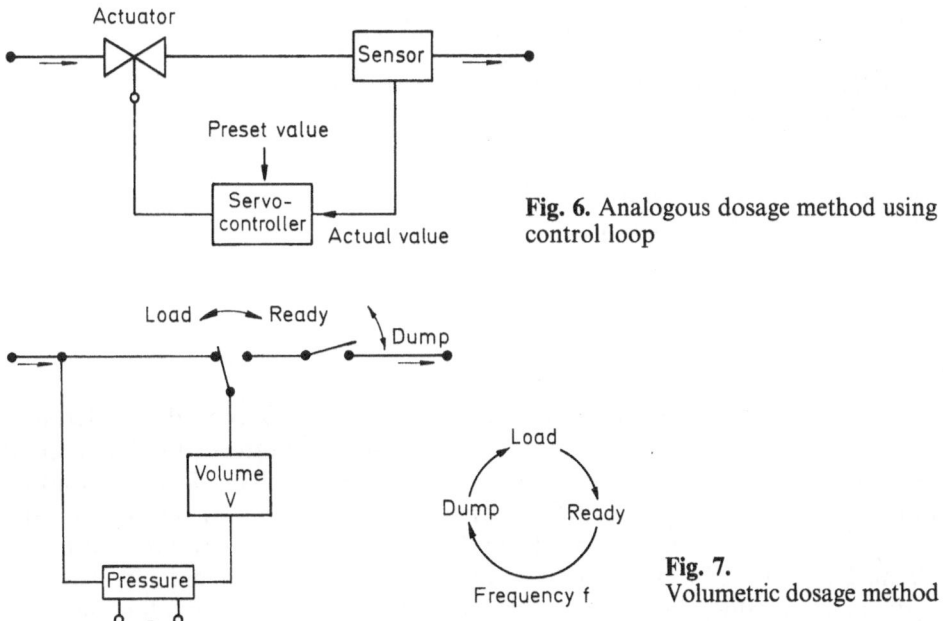

Fig. 6. Analogous dosage method using control loop

Fig. 7.
Volumetric dosage method

Flow measurement in the fresh gas tube for a closed system must always be seen in connection with the dosage method used. This leads to additional requirements. The usual dosage method is analogous dosage, as is illustrated in Fig. 6. The servo-controller always sets the actuator in such a way that the preset value and the actual value agree. The precision of the metering depends on all components of this closed control loop, so that in this case the sensor must be more exact and faster than set forth at the beginning. Thus, in this case, only the hot wire anemometer would be suitable.

An alternative metering method is shown in Fig. 7. The metering is volumetric and the flow follows from $Q_v = V \times p \times f$. The digital character of this kind of metering lends itself very well to the use of a microprocessor, which allows a very exact and stable calibration.

Here it must be pointed out expressly that with this kind of metering a flow measurement, for example with the low flow tubes as shown in Fig. 5, is not necessary. The question as to whether flow meters should be used as a controlling device has not yet been settled.

In conclusion it can be said that present-day technology makes it possible to provide suitable devices for dosage and flow measurement of fresh gas even for completely closed anaesthesia systems.

The Biological Importance of Adequate Oxygen Supply: On-line Measurement of Oxygen and Carbon Dioxide

W. Erdmann, N. S. Faithfull, A. Trouwborst, W. Rating, R. M. Schepp, and R. Droh

Without adequate oxygen supply human life is impaired. Of central consideration is the cell, whose oxygen supply occurs by diffusion from the extracellular space into the intracellular space. Oxygen consumption in the cell results in the production of energy and release of CO_2, which then diffuses out of the cell into the extracellular space. While intracellular PO_2 is maintained fairly constant below 10 mm Hg and PCO_2 below 47 mm Hg, extracellular partial pressures show a wide range of values dependent on the geometric position of the cell in question – in relation to the supplying capillary mesh work. The intracellular gas tensions are maintained at a constant value by changes of the diffusion properties of the cell membrane (Fig. 1). It is not known what the partial pressures are in the nucleus of the cell and current knowledge focuses on the cell in its entirety.

Adequate tissue oxygen supply is dependent upon different parameters:

– Oxygen uptake in the lungs (FIO_2, ventilation/distribution/perfusion ratio)
– Oxygen transport via the blood (O_2 transport capacity/cardiac output)
– Tissue perfusion and distribution (capillary perfusion pressure gradient)
– Oxygen diffusion across the capillary wall (through the interstitial space and cell membrane)
– Oxygen consumption of the cells

Oxygen uptake in the lungs is dependent upon the alveolar-capillary diffusion capacity, the inspired oxygen concentration (FIO_2) and the distribution of the ventilated gas in relation to the perfusion of blood in the various parts of the lungs. Oxygen transport capacity (the amount of oxygen that can be carried in 100 ml of blood at a certain partial pressure) is a very important determinant of oxygen

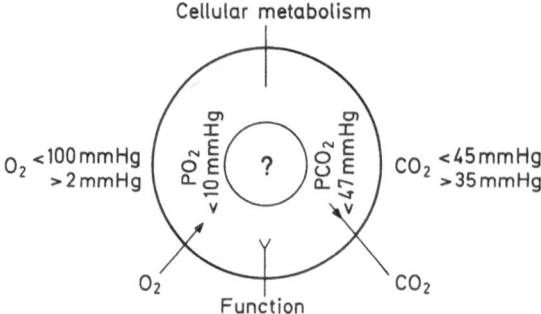

Fig. 1

delivery at the tissues. Changes in the viscosity of the blood (Gosslinga 1982) and the consequent changes in systemic peripheral resistance may influence cardiac output – the amount of blood that is pumped through the tissue per unit of time. When considering tissue perfusion, one should not only look at tissue perfusion in its total, but one should also consider whether perfusion through the tissue is homogenous or inhomogenous. A sample of tissue may be considered to have "adequate tissue perfusion" but some tissue areas may have luxury perfusion and in other areas there may be no perfusion at all. Thus, distribution of perfusion must be efficient. Tissue perfusion is determined by the capillary perfusion pressure gradient of each capillary and is dependent upon the diameter of that vessel and the viscosity of the blood perfusing it. From the capillaries, oxygen has to diffuse across the capillary wall, through the interstitial space and across the cell membrane into the cell. Normally, oxygen supply to the cell is continuously regulated according to the changing consumption of the cells. Higher consumption will be met with a corresponding increase in oxygen supply.

Microphysiological Methods for Determination of the Various Parameters of Oxygen Supply

Modern microphysiology operates with very small electrodes which detect different parameters in the tissue. The basis of the electrode system is a drawn out glass capillary of 50 µm diameter. On this drawn out glass capillary are taped gold wires covered by glass, each 5 µm in diameter. Six gold glass wires taped at equal distances around this glass capillary, with an additional one in the middle, will create a distribution of the gold glass wires according to the geometrics of a hexagon (Fig. 2). The measuring tip in the middle is as far from each angle of the hexagon as is the distance between different angles. Thus a two-dimensional equal distribution of the electrodes is achieved in which the electrodes are capable of simultaneously measuring PO_2, tissue perfusion (using hydrogen clearance techniques) and action potentials. By moving this multi-electrode forward we can then deter-

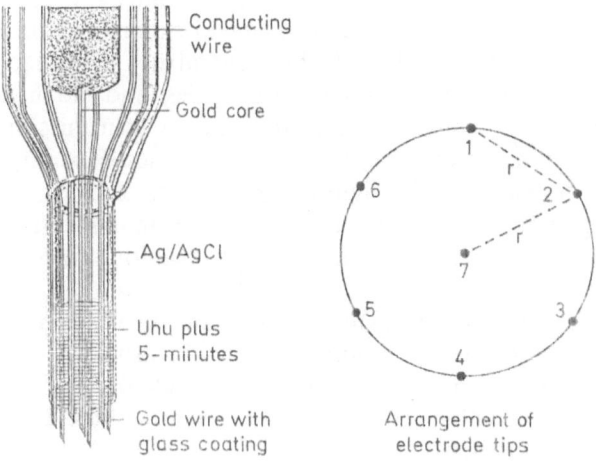

Conducting wire

Gold core

Ag/AgCl

Uhu plus 5-minutes

Gold wire with glass coating

Arrangement of electrode tips

Fig. 2

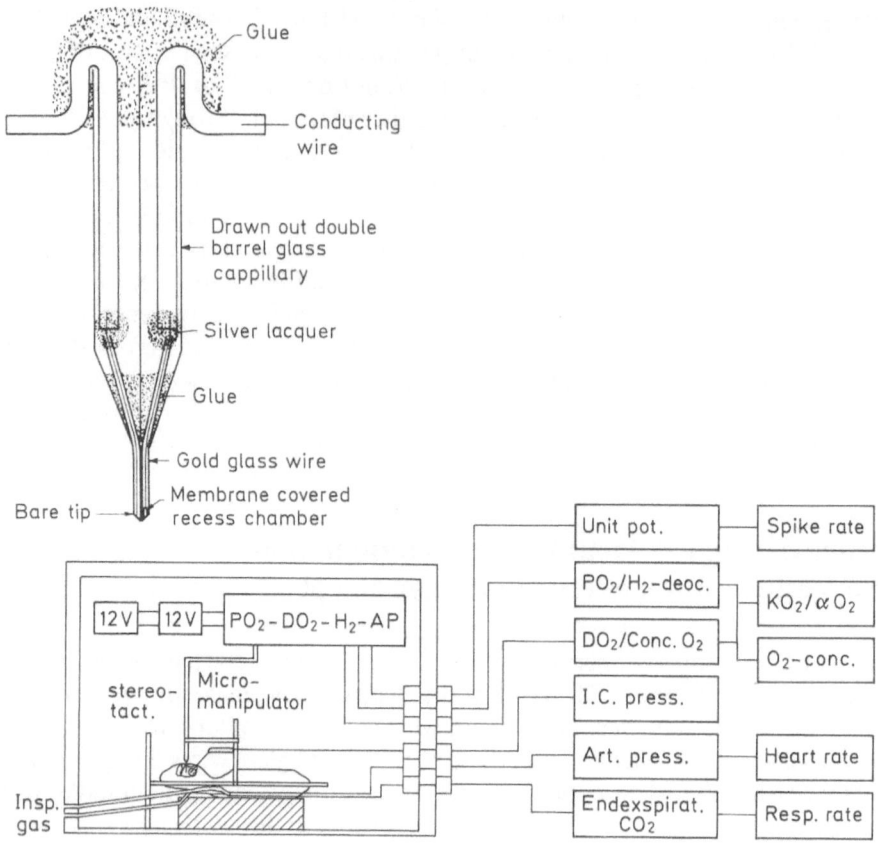

Fig. 3

mine, three dimensionally, the various parameters for tissue oxygen supply in the tissue cylinder and also have some insight into the functional responses (Erdmann et al. 1973; Erdmann 1981). It is important to mention that the 5 μm tip electrodes (1/5000 of a millimetre) give very low currents that have to be amplified, although the electromotor force is also very low. Thus, much noise disturbance is also involved. In order to avoid this noise disturbance, radio waste for example, experiments have to be performed in a double-walled shielded chamber – a Faraday cage. The amplifiers receive their electrical supply from batteries, and the measured values are transmitted to the recording apparatus via light transmission. This consists of a photo emitting diode emitting light at a frequency depending on the voltage which is to bei transferred. The light frequency is received on the other side by a photo receiving diode and modulated again to the original voltage. Thus, on-line recording is possible without feedback noise transmitted into the cage (Fig. 3, lower section).

If oxygen diffusion coefficients are to be measured at the same time, a double-barrelled electrode has to be constructed (Fig. 3, upper section) consisting of a micro tip with a membrane covered recessed chamber (for measurement of PO_2, perfusion and action potentials) and a blank electrode for measurement of the diffusion coefficient (Erdmann and Krell 1976).

18

Characteristics of Oxygen Supply in Neural Tissue

Maintenance of the normal oxygen supply of the brain cortex is one of the most important tasks facing the anaesthetist. What however is "normal"? When a microelectrode is driven step-wise through cerebral tissue (Fig. 4), a characteristic tissue PO_2 profile can be recorded. One can observe high peaks of PO_2 values and very low deep valleys (very low values of tissue PO_2) and then again high, narrow peaks. The PO_2 that is measured depends on the position of the electrode tip relative to the supplying capillaries.

So many values are registered during this sort of experiment that it is impossible to perform statistical analysis by hand and on-line computerisation has to be introduced, especially since multi micro tip measurements are frequently performed. The statistical analysis of distribution of the PO_2 values reveals that about 50% of the tissue PO_2 values in the brain cortex are below 5 mmHg (Fig. 5). The simultaneous recording of action potentials (neuronal unit potentials) demonstrate that under these supply conditions the cells are functioning adequately, even in those areas where, under normal conditions, very low PO_2 values are measured.

Calculation of the distribution of the tissue PO_2 values in the Krogh cylinder system, on the basis of our measurements, reveals numbers of cells far away from the supply capillaries. Indeed, in the venous part of the capillary mesh work, many cells are on the verge of losing their function in the presence of the low PO_2 values that are present. It has to be appreciated that minor disturbances of tissue oxygen supply will easily disturb function and survival of these cells in the "lethal corner" of the capillary mesh work. It should be noted that the above-mentioned measurements were "blind" and it was not known where the electrode was actually positioned at each point of measurement. More information is needed concerning the function of neuronal cells under various supply conditions in the various parts of the supplying system.

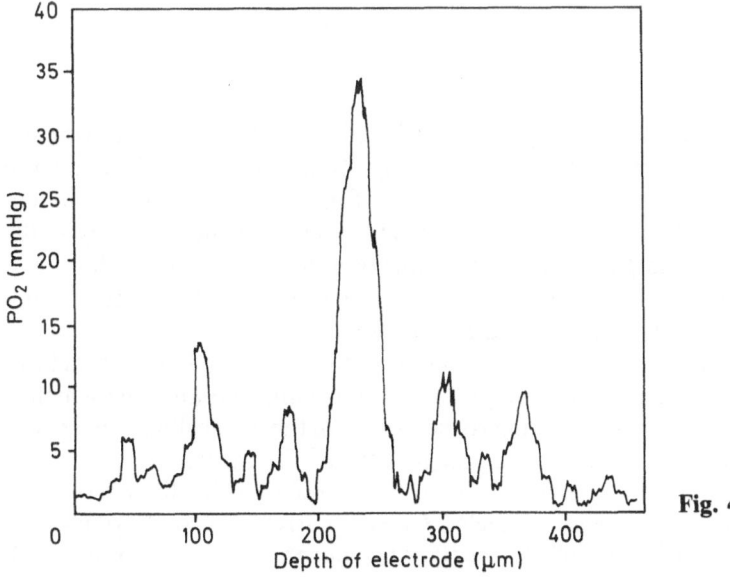

Fig. 4

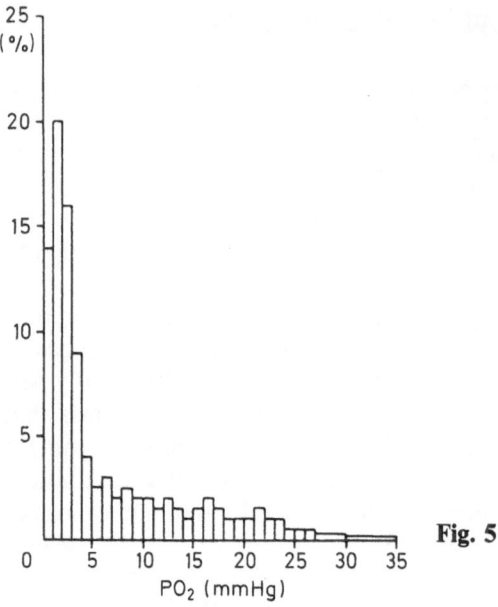

Fig. 5

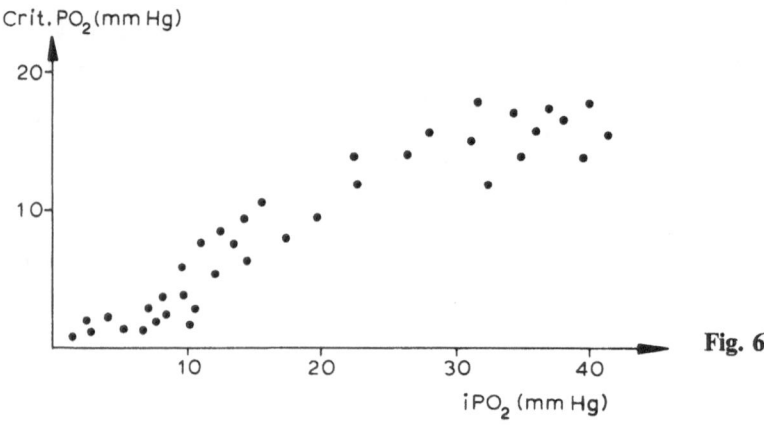

Fig. 6

Simultaneous recording of action potentials and tissue PO_2 values during decrease of oxygen supply by gradual decrease of FIO_2 have given new insights into PO_2 dependent cellular function. Cells, usually functioning in a high PO_2 environment, ceased cell function at 15–20 mm Hg; other cells in originally badly supplied areas can stand PO_2 values of 2 mm Hg and less, remaining fully and normally active. Thus, neuronal cells seem to have undergone environmental adaptation to the varying supply conditions (Fig. 6).

In order to obtain more insight into the above remarkable findings, a functioning neuron had to be found with a high oxygen consumption and one which is large enough to allow placement of micro electrodes in the intracellular space. Experiments were therefore performed in single cell neurons from the abdominal ganglion of Aplysia Californensis (Chen et al. 1978), a snail which lives in the Californian sea near the coast of Los Angeles. Coastal currents keep the water cool and at a constant temperature of 16°–17 °C. The neurons of the ganglion, which

20

have a diameter of ½–1 mm, are spontaneously active and are normally supplied with oxygen by diffusion from the endolymph filled open space of the abdomen.

The neuronally active pacemaker cells are dissected out from the ganglion and placed into a chamber filled with saline solution. Extracellular partial pressures of oxygen can be changed by altering the concentration of oxygen in the gas flowing over and equilibrating with the saline covering the cells. When the oxygen microelectrode approaches the cell membrane, oxygen partial pressure decreases due to consumption diffusion of oxygen into the cell. When crossing the cell membrane a steep gradient of oxygen partial pressure is revealed, from the outside of the cell to the inside, while intra-cellularly there is a fairly homogenous PO_2. Statistical analysis of more than 100 cells of Aplysia Californensis have shown intracellular PO_2's of 5 mm Hg PO_2 concommitent with active cellular function, while extracellular PO_2's of 20 mm Hg PO_2 are normal in the living animal.

Changes of extracellular PO_2 to 40 mm Hg are followed by small increases in the intracellular PO_2. This is then regulated back to the original 5 mm Hg PO_2 within a period of about 1 min. This phenomenon suggests the presence of an autoregulative capacity of the cell membrane. Decrease of extracellular PO_2 to 10 mm Hg PO_2 is followed by a short decrease of the intracellular PO_2 which is immediately counter-regulated to 5 mm Hg. The above findings would suggest that the diffusion resistance of the cell membrane has the capacity of change according to the needs of the cell, to keep the intracellular PO_2 stable (Erdmann and Faithfull 1982). When the extracellular PO_2 is decreased far below 10 mm Hg, intracellular PO_2 is no longer maintained at normal values and even slight permanent decrease to intracellular PO_2 produce an immediate disturbance in cellular spike pattern. After increases of extracellular PO_2 above 50 mm Hg, autoregulatory mechanisms no longer function and the action potential pattern changes to a pattern with epileptic burst-like characteristics.

The above hyperoxic disturbances of the spike rate patterns are assumed to be a sign of oxygen toxicity. These experiments show quite clearly that even slight increases of intracellular PO_2 can have potentially dangerous consequences. Oxygen toxicity affects all tissues of the body and is thought to be one of the major factors leading to "shock" lung. These findings have gained special interest since fluorocarbon containing plasma substitutes seem to have good oxygen transport capacity. The only drawback is that, during treatment, high inspiratory oxygen concentrations are needed (Erdmann et al. 1982).

The Use of Drugs in Brain Protection

Oxygen supply to the cells can be decisively changed by the drugs used in anaesthesia. In addition, various drugs have different actions on cellular metabolism as well as on the micro-perfusional characteristics of the tissue (Erdmann and Kunke 1973). Thus, knowledge of the characteristics of new drugs should also include knowledge about its properties in relation of cellular oxygen supply and consumption.

When the carotid artery of a Mongolian gerbil is occluded, PO_2 in the brain cortex falls to practically zero, in less than 1 s. This is due to the high oxygen consumption of the tissue. This method, using intermittent short-term occlusion of the carotid artery can be used to obtain some insight into the metabolic effects

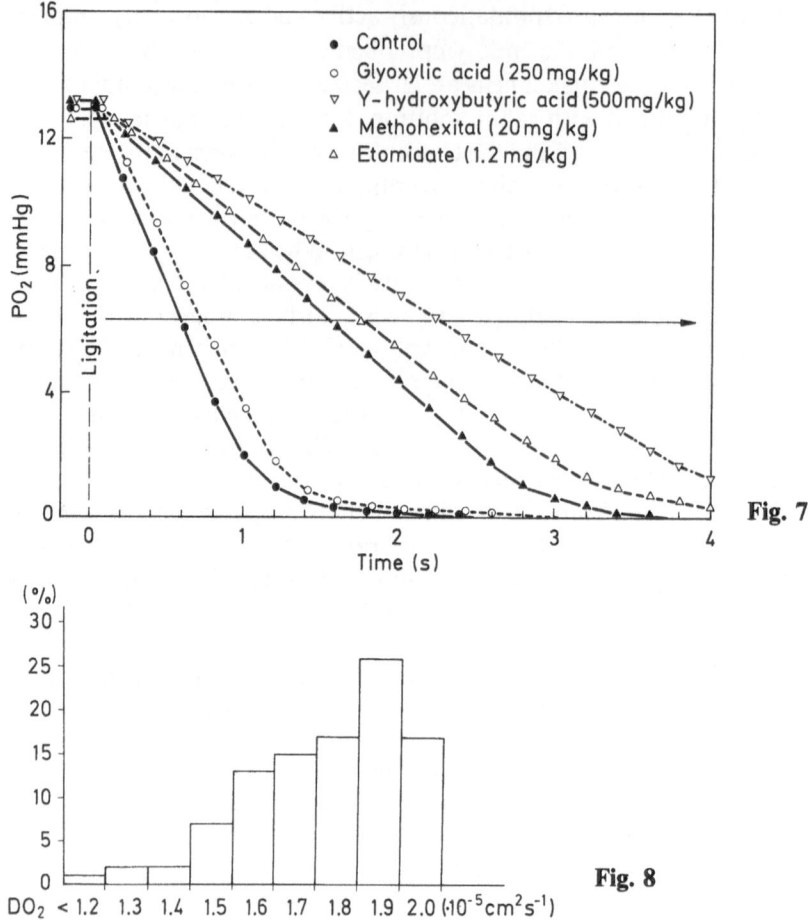

Fig. 7

Fig. 8

of anaesthetic drugs. For example, in an animal treated with glyoxylic acid, the oxygen disappearence curve is not altered decisively. However, injection of barbiturates definitely decreases the oxygen disappearence rate and thus demonstrates that oxygen consumption is very much decreased. Even more striking results are obtained in the presence of gamma-hydroxybutyric acid (Erdmann 1981; Erdmann and Faithfull 1982). Therefore, it is hard to understand why well-known international groups working in concerted action to study brain protection used barbiturates and not gamma-hydroxybutyric acid (Fig. 7).

Another decisive parameter defining cellular oxygen supply is the interstitial oxygen diffusion coefficient. Decrease of the oxygen diffusion coefficient will severely interfere with normal oxygen supply (Erdmann and Faithfull 1982, Morawetz et al. 1978). The oxygen diffusion coefficients in normal neuronal tissue show a wide range of values with a peak at $1.9 \times 10^{-5} \, cm^2/s$ (Fig. 8). There is a close relationship between oxygen diffusion coefficients and normal tissue PO_2. In those areas poorly supplied with oxygen (low PO_2 values) the oxygen diffusion coefficient is rather high, probably as a compensation for the low level of oxygen availability. In the arterial part of the capillary mesh work, with high oxygen partial pressures, the oxygen diffusion coefficient is quite low, probably another regulatory mechanism to avoid hyperoxia of the cells. On the other hand, after

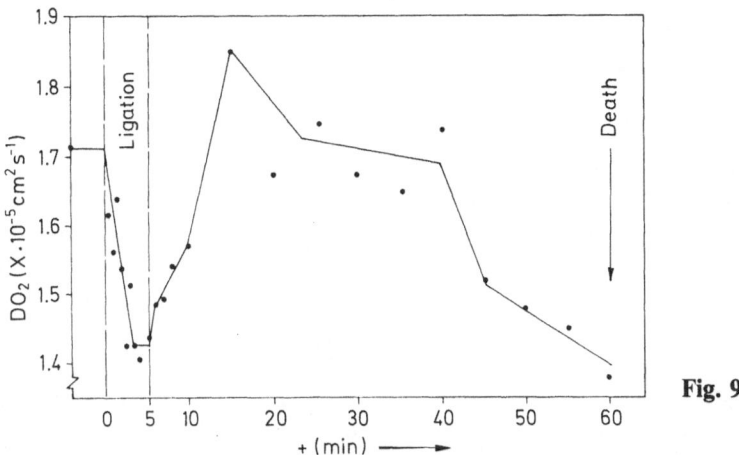

Fig. 9

a short-term hypoxic episode produced by 5 min ligation of both carotid arteries, a short-term recovery of oxygen diffusion coefficient occurs, followed by a decrease to low values. In a number of cases death occurs (Fig. 9), even though tissue perfusion and capillary oxygenation has been restored. In the experimental situation these frequently observed deleterious effects of short-term hypoxic episodes can be reversed by the administration of glucosaminoxidases (hyaluronidase) (Erdmann 1981; Erdmann and Faithfull 1982).

The Role of Carbon Dioxide in Tissue Oxygenation

In the regulation of perfusion in our tissue model, PCO_2 plays a major role in regulating the perfusion of the tissue. This is effected by changes in the wall tension of the arteries and the larger arterioles.

Increase of wall tension means constriction of these vessels concommitant with increase of flow resistance and a larger than normal pressure drop between the arteries and the arterial part of the capillary – normally, the mean arterial pressure of 100 mm Hg decreases to a mean arterial capillary pressure of 60 mm Hg. When the inflow pressure at the arterial end of the capillary is decreased below 60 mm Hg, perfusion is severely reduced. This is a consequence of the decrease of capillary perfusion pressure (mean pressure at the arterial end of the capillary minus mean pressure at the venous part of the capillary). This perfusion pressure is forcing blood through the capillaries and squeezing erythrocytes through the relatively narrower capillary system. The pressure gradient has, furthermore, to be seen in relation to the length of the capillary which, in the cerebral cortex, is normally about 30 μm long. When the capillary length is increased, for example due to partial obstruction of capillaries and collateral capillary perfusion, the perfusion pressure gradient has to be increased to supply adequate perfusion. Thus, the arterial PCO_2 is a decisive regulator of capillary inflow pressure and thus perfusion pressure gradient.

During measurement of tissue PO_2 in different parts of the intercapillary mesh work, it can be seen that variations in arterial PCO_2 between 30 and 42 mm Hg do not alter tissue PO_2, either in well supplied or in badly supplied areas. Increasing the PCO_2 above 42 mm Hg, (corresponding to an end expiratory CO_2 of 6 volumes %) leads to an increase of capillary perfusion due to an increased perfusion pres-

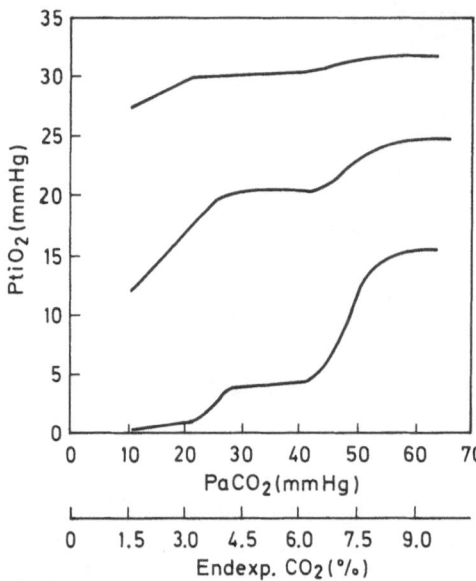

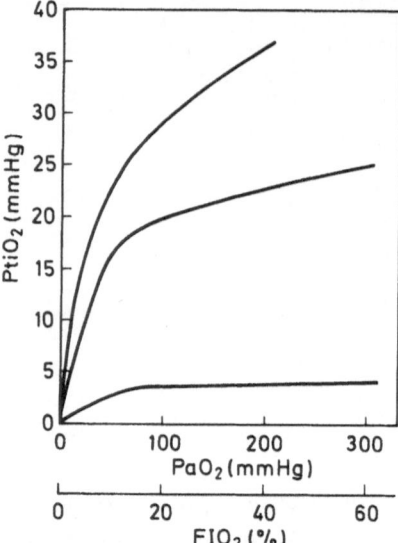

Fig. 10   Fig. 11

sure gradient. This affects tissue PO_2 in the badly supplied area (Fig. 10, lower trace : initial PO_2 3.75 mm Hg) to a greater extent than it does the PO_2 in well supplied areas (Fig. 10, upper trace : initial PO_2 30 mm Hg) (Erdmann and Metzger 1971). Tissue perfusion is maximally increased at an end expiratory CO_2 of around 8%.

A more severe effect of changes of arterial PCO_2 is observed during hyperventilation. Decreasing arterial PCO_2 below 25 mm Hg decreases tissue PO_2 in the venous part of the capillary mesh work to values less than 50% of normocapnic values. At 20 mm Hg PCO_2 almost complete anoxia exists in these badly supplied areas. In contrast, in areas in the capillary mesh work well supplied with oxygen and with an initial PO_2 values of 30 mm Hg and more, hypocapnia produced by hyperventilation has much less effect. These areas are probably the only ones to survive long term hyperventilation. Unfortunately, hyperventilation is still widely applied and there rarely exists monitoring possibilities for on-line determination of end expiratory CO_2.

Changes of inspiratory oxygen concentration above the normal value of 21% with a consequent increase of arterial PO_2 into the hyperoxic range has little effect on the PO_2 of tissue areas in the "lethal corners" while it has a tremendous effect on those areas which are already well supplied with a PO_2, of 30 mm Hg, ore more (Erdmann and Metzger 1971). Hypoxia on the other hand affects all areas of the tissue cylinder but has a greater effect on those areas in the arterial part of the capillary mesh work as compared with those in the venous part (Fig. 11).

Conclusions

Based on the information presented in this paper, clinical goals should be aimed at achieving adequate on-line determination of parameters connected with tissue supply of oxygen.

These aims should be:
- On-line determination and feedback control of inspiratory oxygen concentration
- On-line calculation of oxygen consumption
- Non-invasive intermittent determination of ventilation perfusion ratio of the lungs
- Non-invasive intermittent determination of cardiac output

Continuous monitoring of end expiratory CO_2 is already mandatory in some countries (e. g. The Netherlands), but the real significance of this measurement does not yet seem to have been widely accepted throughout the world. On-line determination of CO_2 production is necessary to determine the respiratory quotient and, furthermore, the pattern of the respiratory CO_2 can give pertinent information about acute pathophysiological changes (Smalhout and Kalendar 1975).

Before application of new anaesthetic drugs it should be mandatory to investigate the effects of the drug on the physiology of the micro areas, with special reference to cellular oxygen metabolism, cellular oxygen supply and oxygen diffusion parameters in the lung and tissue. In addition, effects on the micro circulation and on blood viscosity should be studied. The investigations are not usually performed, although intensive investigation of the cardiovascular effects, of the effects on the endocrine system, on neurophysiological behavioural effects and the effects on ventilation are routinely performed.

Proposition

Each day there are most probably billions of human neurons killed due to improper administration of anaesthetics and anaesthesia, especially with uncontrolled and improper ventilation. This is due to lack of knowledge concerning the O_2–CO_2 inter-relationship, the physical characteristics of the transporting medium and the impact of ventilation/perfusion/diffusion parameters on the gas exchange in total.

References

Chen CF, Erdmann W, Halsey JH (1978) The sensitivity of aplysia giant neurons to changes in extracellular and intracellular PO_2. IN: Silver IA et al (eds) Oxygen Transport to Tissue III. Plenum, New York, pp 780–786

Erdmann W (1981) Continuous on-line acquisition of tissue PO_2, tissue perfusion, oxygen diffusion parameters and neuronal activity. In: Kimmich HP (ed) Monitoring of vital parameters during extracorporeal circulation. Karger, Basel, pp 73–82

Erdmann W, Faithfull NS (1982) The disturbance of cellular oxygen supply in the post hypoxic period. IN: Wauquier A et al (eds) Protection of tissues against hypoxia. Elsevier/North Holland Biomedical, Amsterdam, pp 183–197

Erdmann W, Krell W (1976) Measurement of diffusion parameters with noble metal electrodes. In: Grote J, Reneau D, Thews G (eds). Oxygen transport to tissue II. Plenum, New York, pp 225–228

Erdmann W, Kunke S (1973) Changes of oxygen supply to the tissue following intravenous application of anesthetic drugs. In: Bicher I, Bruley D (eds) Oxygen Transport to Tissue. Plenum, New York, pp 261–269

Erdmann W, Metzger H (1971) Oxygen tension in microareas of the brain. Steady states of oxygen tension at arterial hyperoxia and with hypercapnia-increased CBF. In: Poedlesch I (ed). 6th European conference on microcirculation, Aalberg 1970. Karger, Basel, pp 303–307

Erdmann W, Kunke St, Krell W (1973) Tissue PO_2 and cell function: an experimental study with multimicroelectrodes in the rat brain. In: Kessler W (ed) Oxygen supply. Urban and Schwarzenberg, München

Erdmann W, Frey R, Madjidi A, Beisbarth H (1982) Oxygen-carrying v. non oxygen-carrying colloidal blood substitutes in shock. Injury 14:70–74

Gosslinga H (1982) The viscosity of blood. An experimental study into the effects of alterations in blood viscosity during shock. PhD thesis, University of Utrecht. Utrecht Drukkerij, Elinkwijk, p 43

Morawetz R, Strong E, Clark DK, Erdmann W (1978) Effects of ischemia on the oxygen diffusion coefficients in the brain cortex. IN: Silver IV et al (eds) Oxygen Transport to Tissue III; Plenum, New York, pp 629–632

Smalhout B, Kalender Z (1975) Textbook of capnography, vol 1. Kerckebosche, Zeist, Holland

Principles of Measurement of Anaesthetic Agent Concentrations in Closed-Circuit Systems

H.-D. Hattendorff and C. F. Wallroth

The measurement of concentrations of anaesthetic agent and nitrous oxide in the breathing circuit may be carried out at three sites in the system: measurement in the inspiratory limb allows at least control of the gas actually led to the patient, while measurement between Y-piece and patient or in the expiratory limb of the circuit will give important information on expiratory concentration values. Ideally, the end-expiratory value is measured.

The anaesthetic agents halothane, enflurane and isoflurane in particular are discussed here. Criteria according to which possible methods of anaesthetic agent and nitrous oxide measurement may be evaluated are:

1. Selectivity. Halothane, enflurane or isoflurane are to be measured in the presence of the residual gas, which is composed of changing proportions of O_2, N_2O, H_2O, CO_2 and possibly N_2. Similarly, N_2O is to be measured, among others, independently of the anaesthetic agent.
2. Measuring range. The range should be O to max. 10 vol.% anaesthetic and O to max. 100 vol.% N_2O.
3. Measuring accuracy. The measuring accuracy to be demanded depends on the actual purpose. Absolute measuring accuracies in the range of a few per cent of reading should be generally sufficient. It is important, however, that the required measuring accuracy be maintained in every working condition.
4. Response time. Maximum requirement would be the resolution-in-time of the concentration course within individual breaths; for that response times (t_{90}) of the order of 100 ms are required. For measurement of mean concentration values, response times of several seconds or several tens of seconds are sufficient.
5. Suction rate. Methods in which an amount of gas is sucked from the breathing circuit and led to the actual measuring device (and not led back into the breathing circuit) are only suitable, particularly for the closed system, if the volume sucked per time unit is small enough, smaller than approximately 10 ml/min.
6. Delay time in suction methods. Time from sample-taking to display of the measured value should ideally be shorter than the time taken by several breaths; in any case, it should be short enough for the displayed value still to be relevant.
7. Stability. The error resulting from zero or gain drift should be small enough to maintain the measuring accuracy during at least one period of operation.
8. Possibility of sterilization

9. Handling
10. Expense.

Potential methods of measuring anaesthetic agent and nitrous oxide concentrations include:

Silicone rubber membrane   Infra-red absorption
Quartz oscillator   Micro-wave absorption
Mass spectroscopy   Infra-red fluorescence
Gas chromatography   Radioactive tracing
Raman spectroscopy   Refractometry.
Ultraviolett absorption

Essentially, methods are considered here which are suitable for continuous measurement (exception: gas chromatography). In the method listed first, expansion and change in elasticity of a silicone rubber membrane, resulting from absorption of anaesthetic agent, are utilized as measuring effect. The measuring method is slow and is subject to cross-sensitivity against nitrous oxide and water vapour and thus is hardly suitable for measurement of anaesthetic agent concentration in the breathing circuit. The fact that a silicone rubber layer, for instance, absorbs anaesthetic agents can be utilized in a different way: if this layer is applied to a quartz resonator, the additional mass absorption on the quartz, resulting from uptake of anaesthetic, leads to a change in the oscillation frequency. This frequency change is a measure of the concentration of the anaesthetic agent. However, this method is also subject to cross-sensitivity against nitrous oxide and particularly water vapour (an absorption layer insensitive to water vapour has so far not been discovered). Therefore, this method of measuring anaesthetic agent in the breathing circuit also has only a limited application (Cooper et al. 1981; Hayes et al. 1983).

The basis for the refraction method are the differing refraction indices of the anaesthetic vapour and the carrier gas. In principle, only binary gas mixtures can be measured, thus the method is not suited for measurement of anaesthetic vapour concentration at changing residual gas composition without additional effort; furthermore, the measuring time is too long. It should be emphasized, however, that the calibration of corresponding instruments (interferometer) will remain stable for years and that these instruments are suited for precision measurement of binary gas mixtures. They are therefore used as reference for calibration of anaesthetic agent vaporizers or for calibration of other measuring methods.

If the absorption of anaesthetic vapours in the ultraviolet spectral region is used for concentration measurement, the sample must be sucked off and must not be reintroduced into the patient system, because the ultraviolet radiation may possibly result in formation of toxic fragments of the anaesthetic. The estimated suction rate necessary for a time-resolved measurement is too high for the closed system. However, it is possible to measure a mean concentration, obtained over some time, at a reduced suction rate (Nunn 1979). Via ultraviolet absorption it is relatively simple to measure halothane selectively against enflurane and isoflurane.

To our knowledge the absorption of halothane, enflurane, and isoflurane in the microwave region has not been closely examined. If one looks at other hydrocarbons (see Zeil 1978), however, the expected absorption lengths for anaesthetic va-

28

pours are too long. It should also be examined whether under atmospheric pressure the anaesthetic can still be measured selectively against the residual gas.

As far as we know, the fluorescence of anaesthetics in the infra-red spectral region has not been examined up to now either, but the relatively high expense (a strong infra-red source, e.g. a laser, is necessary; see Robinson and Dake 1973) does not seem justified for breathing gas measurement in view of advantages hardly expected in comparison with other measuring methods.

Gas measurement via tracing of radioactive isotopes (see Smidt et al. 1976) may be difficult in practice and would be suitable for very special applications only – perhaps measurement over particular lung regions.

In mass spectroscopy, a gas volume is continuously sucked from the breathing circuit through a capillary tube. Towards the end of the capillary tube, where the gas pressure has decreased to approximately 1 Torr, the gas molecules may, via a porous gold foil, for instance, diffuse into the high vacuum chamber of the mass spectrometer. There they are ionized by an electron beam accelerated in an electric field and spatially separated in a magnetic field perpendicular to the flight direction, according to their charge/mass ratio; at the impact point the ion current is measured, this being a measure of the partial pressure of the amount of the relevant component present in the breathing gas. Instead of the spatial separation in the magnetic field, it can be provided by means of appropriate electric field – utilizing only one ion collector – that only ion of a particular charge/mass ratio will reach the collector (quadrupole mass spectrometer); by varying the fields during a measuring cycle the current of ions of different charge/mass ratios can be measured in timed sequence. The third variation is the impulse transit time mass spectrometer, with which the ions, accelerated in pulses in an electric field, strike a collector in timed sequence according to their charge/mass ratio.

In ionization of the gas molecules by means of the electron beam, also (charged) fragments of the original molecules are produced. Cross-sensitivities which at first are present at equal mass number – e.g. N_2O/CO_2 – may be bypassed, either by measuring such fragments or by measuring different naturally occurring isotopes and possibly evaluating signals of different channels (one channel corresponding to one charge/mass ratio) (see Smidt et al. 1976).

With the mass spectrometer not only anaesthetic agents and nitrous oxide but all components in the breathing gas can be measured (see Sodal and Swanson 1979). However, measurement of water vapour, because of adsorption of water vapour at the capillary tube walls, poses a problem; this effect may even affect the measurement of the other gas components, or extend the response time. Sodal and Swanson (1980) describe a special inlet system for the breathing gas measurement which foregoes use of the capillary tube; thus also measurement of water vapour is made possible.

For mass spectrometers response times (t_{90}) of 100–500 ms are stated and delay times of 200–600 ms at capillary tube lengths of 1–2 m. The suction rate is 10–60 ml/min. Suction may be carried out at the Y-piece or elsewhere. The measuring accuracy of mass spectrometers in principle is fully sufficient for breathing gas measurement. The measuring system is complex and means considerable expense.

Like mass spectrocopy, gas chromatography is also a separation process. In gas chromatography, the sample to be analysed, typically approximately 1 ml, is added to a carrier gas, e.g. He or N_2. The carrier gas together with the sample flows through a column, the interior wall of which is lined with a suitable material. As the individual gas components contained in the sample have differing affinities to this coating material, the components will successively leave the column together with the carrier gas. The separating column is connected to a detector. Among others, the thermal conductivity detector, the flame ionization detector and the electron capture detector are used as detectors. With the thermal conductivity detector the differing thermal conductivities of the carrier gas and the component to be detected are utilized. In the flame ionization detector the gas mixture to be examined is burned with hydrogen and air and the resulting ionization current is measured; for the purpose the burner nozzle is surrounded by a cylindric electrode which, in relation to the nozzle, has a potential of approximately $+100$ V. This detector responds to organic substances, thus especially to anaesthetics. And finally, the electron capture detector consists of a pair of coaxial electrodes. The positive central electrode is surrounded by a hollow cylindrical zirconium cathode which is impregnated with a 20 mC tritium-β emitter; the potential difference between anode and cathode is approximately 10 V. Halogens present in the space between the electrodes capture some of the electrons emitted from the β-source; thus the electron current between anode and cathode will decrease. Particularly the electron capture detector is sensitive to halogenated hydrocarbons, such as halothane, enflurance, and isoflurane.

Instead of one separating column, several columns may be connected in parallel. The columns may have differing lengths and may be lined with different materials (see Hill 1970). Thus all components of a breathing gas mixture may be measured.

Via gas chromatography only a discontinuous measurement is possible, with samples taken from the breathing gas from time to time. The time after which the result is available essentially depends on the separating time in the column and varies between 30 s and 5 min. An automatic control to take only end-expiratory samples would be complex. Measuring accuracy necessary for breathing gas measurement can be reached with gas chromatography and also sufficient stability could be obtained.

In principle, the flame ionization detector can be applied directly – without preceding separation – to measurement of anaesthetic vapours in the breathing gas (Osinga and Heuler 1979). Here, gas is sucked from the breathing circuit via a capillary tube (3 ml/min). However, problems with absorption of water vapour in the capillary tube have been reported, as well as interference in the measurement by changing proportions of N_2O and O_2 (change in the viscosity) in the residual gas. Furthermore, application of a flame ionization detector in the operating theatre will necessitate additional safety measures (flame, hydrogen).

Raman spectroscopy utilizes inelastic scattering of light by the gas molecules. The frequency of the scattered light is, in relation to the entering light, shifted upward or downward by an amount which corresponds to the change in the vibration and/or the rotation state of the molecule. The intensity of the scattered light is measured selectively, at the characteristic frequency for the respective kind of molecule (possibly at several frequencies characteristic for the kind of molecule). The intensity of the scattered light is, among others, proportional to the concentration, more precisely, to the density of molecules of the respective gas component. Measurement of the scattered light at different frequencies which are specific for the respective components makes possible the determination of all breathing gas components, including anaesthetic agents and water vapour (but not inert gases). The Raman effect depends on the change of polarizability which is connected with the transition between two vibration and/or rotation states of the molecule. On the other hand, infra-red absorption is observed at a transition between two vibration levels (possibly coupled with a transition between two rotation levels) of the molecule, if the change of the vibration state is connected with a change in the electric dipole moment. Thus O_2 and N_2 for example, which do not show infra-red absorption, can be measured by Raman spectroscopy.

The measurement of breathing gas by Raman spectroscopy (see Albrecht and Schaldach 1975) in principle may be performed directly in the breathing circuit. A laser in the visible spectral region is best used as excitation source, preferably in the blue region. Measurement accuracies of a few per cent of reading and response times down to 100 ms may be obtained. However, up to now, breathing gas measurement utilizing the Raman effect has not been exclusively studied. Complexity and expense of the method are comparable with those of mass spectroscopy.

Less expensive although still ambitious enough is the measurement of anaesthetic vapour concentration and nitrous oxide content in the breathing gas via infra-red absorption. The concentration, or more precisely the density, of molecules of the respective component is measured via light attenuation at a wavelength in the infra-red spectral region specific to this gas component. Thus the anaesthetic agent, N_2O and CO_2 for instance, can be measured selectively without interference by water vapour. The infra-red absorption method offers good accuracy, high stability and response times in the order of 100 ms. Measurement may be carried out directly in the breathing circuit, which means without suction of a sample. The method allows design of relatively compact measuring instruments.

When the different methods are compared, infra-red absorption appears to be particularly attractive for measuring anaesthetic agent and nitrous oxide concentration in the breathing circuit, especially in the closed system. Therefore, an instrument has been developed which, by infra-red absorption, measures halothane, enflurane, and isoflurane – the proper sensitivity for the actual anaesthetic agent is selected via front panel keys – and simultaneously nitrous oxide directly in the breathing circuit. The instrument consists of the sensor and the signal-processing unit. The sensor may be placed in the inspiratory or in the expiratory limb of the anaesthesia system. Sensor and signal-processing unit are sterilizable.

The active cell volume is approximately 16 ml. Response time (t_{90}) is about 100 ms. The measuring range is 0 to 10 vol.% anaesthetic and 0 to 100 vol.% N_2O. The absolute measurement accuracy is better than a few per cent of the reading. The measurement is not affected by water vapour or CO_2.

The instrument gives a digital read-out of the anaesthetic agent and nitrous oxide concentrations. For the present, the highest or lowest concentration value (as selected by keys) is displayed, which has been encountered during a time window, adjustable between 1 and 60 s or synchronizable with the respiratory frequency. Independently of this, the current value of the anaesthetic agent and nitrous oxide concentration is available at the analogue output for continuous recording.

To summarize, for anaesthetic agent and nitrous oxide measurement in the breathing circuit, a complex method such as mass spectroscopy may be the method of choice in special cases. However, many measurement problems – among others, measurement in the closed system – may be covered by instruments based on infra-red absorption.

References

Albrecht H, Schaldach M (1975) Ramanspektroskopische Gas-Partialdruckbestimmung für eine kontinuierliche Überwachung der Blut- und Atemgase. Biomed Tech (Berlin) 20:119

Cooper JB, Edmondson JH, Joseph DM, Newbower RS (1981) Piezoelectric sorption anesthetic sensor. IEEE Trans Biomed Eng 28(6):459

Hayes JK, Westenskow DR, Jordan WS (1983) Monitoring anesthetic vapor concentrations using a piezoelectric detector: evaluation of the Engstrom emma. Anesthesiology 59:435

Hill DW (1970) Electronic measurement techniques in anaesthesia and surgery. Butterworths, London

Nunn JF (1979) Monitoring of totally closed systems. In: Aldrete JA, Low HJ, Virtue RW (eds) Low flow and closed system anaesthesia. Grune and Stratton, New York

Osinga R, Heuler R (1979) Ein neues Gerät zur kontinuierlichen Messung von Narkosedämpfen. Prakt Anaesth 14:182

Robinson JW, Dake JD (1973) Parameters governing the intensity of laser induced infrared fluorescence. Spectrosc Lett 6:685

Smidt U, von Nieding G, Löllgen H (1976) Methodische Probleme der Atemgasmessung. Biomed Tech 21(4):102

Sodal JE, Swanson GD (1979) Mass spectrometry: current technology and implications for anesthesia. In: Aldrete JA, Lowe HJ, Virtue RW (eds) Low flow and closed system anesthesia. Grune and Stratton, New York

Sodal E, Swanson GD (1980) Mass spectrometer in patient monitoring. In: Gravenstein JF, Newbower RS, Ream AK, Ty Smith N, Barden J (eds) Essential noninvasive monitoring in anesthesia. Grune and Stratton, New York

Zeil W (1978) Qualitative und quantitative Analysen von Gasgemischen durch Mikrowellenspektroskopie. Fresenius Z Anal Chem 289:1

The Present Status of Nitrous Oxide in Clinical Anaesthesia

J. W. R. McIntyre

Nitrous oxide has been used in clinical anaesthesia for more than a century. Supplementary drugs have usually been added because of the limited central nervous system depressant effects of 70% nitrous oxide (Hornbein et al. 1982) and the oxygen requirements of some patients (Nunn 1969). Altitude augments these limitations (James et al. 1982). In 1978 our interest in reducing operating-room contamination with nitrous oxide led us to obtain experience with closed-circuit or low fresh gas flow techniques and to re-examine our routine use of nitrous oxide for almost every patient who reached the operating-room suite. This reassessment entailed a review of contra-indications to the drug, considering the evidence for our historical belief in certain effects of the drug, and concluding whether alternative drug combinations might be more appropriate.

One important physical characteristic is a degree of solubility that permits entry into closed gas-containing spaces that may exist in the patient. These include a subdural space (Mallamo et al. 1975; Mac Gillivray 1982; Artru 1982), lumbar or cisternal space (Saidman and Eger 1965), the middle ear (Perreault et al. 1981; Casey and Drake-Lee 1982), pneumothorax (Eger and Saidman 1965), and pneumoperitoneum, the intestinal tract (Eger and Saidman 1965; Moens and DeMoor 1981), and emboli (Munson 1971; Steffey et al. 1980). Reports of related morbidity exist and the best-documented is ear injury due to raised middle ear pressure (Owens et al. 1978; Carter and Noval 1981; Srivastava 1980; Man et al. 1980). Controversy regarding tension pneumoencephalos (Friedman et al. 1981, Friedman 1983) concerns the relative times of nitrous oxide administration and the closing of the dura following craniotomy, and the debatable issue regarding gas emboli is whether the hazard of increasing the size of the embolus is outweighed by the advantage of rendering it easier to detect (Munson 1971; Shapiro et al. 1982).

Certain anaesthesia equipment can also be affected by the presence of nitrous oxide – endotracheal tube cuffs (Bernhard et al. 1978; Brandt et al. 1982; Munson et al. 1980; Kopp and Wehmer 1981), ballon-tipped catheters (Kaplan et al. 1981; Du Boulay and Nahrwald 1982), vaporizers (Stoelting 1971; Knill e tal. 1980; Lampert et al. 1982), and oxygen analysers (Piernan et al. 1979; Orchard and Sykes 1980). However, these changes in cncentration delivered from vaporizers and dimensions of air-filled balloons often vary with the particular commercially available item. They can be determined prior to use and except in the case of oxygen or vapour analysis are unlikely to constitute a serious hazard to a patient.

The routine use of nitrous oxide has been based in part on the belief that by reducing the necessary minimal alveolar concentration (MAC) of the inhalational agent necessary to supplement it the patient is spared some of the undesirable effect of the supplemental agent. Occasionally, anaesthetists wish patients to breathe spontaneously and substituting nitrous oxide for an equivalent amount of enflurane vapour has a beneficial effect on ventilation (Lam et al. 1982). This improvement is substantially greater than that achieved with halothane or isoflurane (Hornbein et al. 1969; Eger et al. 1972). Clearly, nitrous oxide makes it easier to achieve adequate spontaneous respiration.

During the past decade the cardiovascular effects of combinations of nitrous oxide with anaesthetic vapours and narcotic analgesics have been studied intensively. In 1971 Bahlman et al., examining the cardiovascular effects of halothane with or without nitrous oxide, concluded that the differences were small, and in elderly or debilitated patients and those receiving sympatholytic medication the small cardiovascular benefit accruing from the combination of nitrous oxide with halothane would disappear (Bahlman et al. 1971; Steffey et al. 1974). Subsequent studies showed that the addition of nitrous oxide during halothane anaesthesia produced significant cardiovascular changes – stimulation or depression – that were variable and dependent on many factors (Hill et al. 1978). The results of a study of the impact of nitrous oxide on the circulation during enflurane anaesthesia provoked a conclusion that the apparent protection afforded by nitrous oxide was so small that the main consideration in using nitrous oxide should be the inspiratoric oxygen concentration (FIO_2) ideal for that patient (Smith et al. 1978), and that under certain circumstances nitrous oxide contributes to enflurane-induced cardiovascular depression (Hanowell et al. 1972). Similar findings during narcotic analgesic nitrous oxide combinations have been reported (Lappas et al. 1975; Lunn et al. 1979).

Studies of baroreflex control in man have shown that nitrous oxide lessens the dose-related depression of this reflex by halothane (Duke and Trosky 1980), but a similar effect could not be found during enflurane anaesthesia (Morton et al. 1980).

A general conclusion from published work about the cardiovascular effects induced directly or indirectly by the addition of nitrous oxide is that they are small, stimulating or depressant. In any particular patient they are difficult to predict because of the numerous interacting factors, which include time, the quantity of volatile agent, the patient's pathophysiology and concomitant medication. Particular attention has been drawn by Schulte-Sasse et al. (1982) to the significant increase in pulmonary vascular resistance occurring in patients with chronic mitral valvular stenosis as opposed to coronary arterial disease, and the undesirability of nitrous oxide in these patients has been confirmed (Davidson and Chinyanga 1982). It is concerning the health of nurses and anaesthesiologists as well as patient welfare that other adverse cellular effects of nitrous oxide have been of interest. However, effects on bone marrow (Anonymous 1978), the mobility (Nunn and Morain 1982) and microbicidal oxidative function of human neutrophils (Welch and Zaccari 1982), and methionine synthetase (Waskell et al. 1982) have been demonstrated. The role of such changes in postoperative morbidity – particularly for surgical procedures lasting 12–24 h – is uncertain at the present time.

From all the foregoing it appears that in a few procedures nitrous oxide is contra-indicated and in many others it offers no particular advantages. The disadvantage is that its presence in the anaesthesia delivery system increases the likelihood of a serious hypoxic episode for the patient.

A simple and versatile alternative to nitrous oxide and oxygen as a vehicle and supplement for volatile halogenated agents is to provide the patient with an air-oxygen mix and approximately 70% MAC of a volatile halogenated agent (Murphy and Hug 1982 a, b). This is supplemented with an intravenous narcotic analgesic or more of the volatile agent as required. We have used enflurane-fentanyl or isoflurane-fentanyl in this manner for a variety of surgical procedures. No problems have occurred in the operating-room and the course in the postanaesthesia recovery-room has been similar to that following other general anaesthetic methods. More sophisticated assessment is yet to be done.

In conclusion, the future trend in general anaesthesia in large and small hospitals will probably be towards the use of low fresh gas flows of air-oxygen mix: a volatilized agent sufficient to produce in the patient a fraction of its MAC, supplemented by a narcotic analgesic delivered by an intravenous route.

References

Anonymous (1978) Nitrous oxide and the bone marrow. Lancet 2(8090):613

Artru AA (1982) Nitrous oxide plays a direct role in the development of tension pneumocephalus intra-operatively. Anesthesiology 57(1):59

Bahlman SH, Eger EI II, Smith NT et al. (1971) The cardiovascular effects of nitrous oxide-halothane anesthesia in man. Anesthesiology 35(3):274–285

Bernhard WN, Jost LC, Turndorff H et al. (1978) Physical characteristics of and rates of nitrous oxide diffusion into tracheal tube cuffs. Anesthesiology 48:413

Brandt L, Pokar H, Renz D et al. (1982) Cuffdruckänderungen durch Lachgasdiffusion. Anaesthetist 31:345

Carter JA, Nofal FM (1981) Tympanic membrane rupture during nitrous oxide anaesthesia. Br J Anaesth 53(2):194

Casey WF, Drake-Lee AB (1982) Nitrous oxide and middle ear pressure. Anaesthesia 37:896–900

Davidson JR, Chinyanga HM (1982) Cardiovascular collapse associated with nitrous oxide anaesthetic: a case report. Can Anaesth Soc J 29:484

DuBoulay PMH, Nahrwold ML (1982) In vivo response of air-filled balloon-tipped catheters to nitrous oxide. Anesthesiology 57:530–532

Duke PC, Trosky S (1980) The effect of halothane with nitrous oxide on baroreflex control of heart rate in man. Can Anesth Soc J 27(6):531–534

Eger EI II, Saidman LJ (1965) Hazards of nitrous oxide anesthesia in bowel obstruction and pneumothorax. Anesthesiology 26:61

Eger EI II, Dolan WM, Stevens WC et al. (1972) Surgical stimulation antagonises the respiratory depression produced by Forane. Anesthesiology 36:544

Friedman G, Norfleet EA, Bedford RF (1987) Discontinuance of nitrous oxide does not prevent tension pneumocephalus. Anesth Analg (Cleve) 60:57–58

Friedman GA (1983) Nitrous oxide and the prevention of tension pneumocephalus after craniotomy. Anesthesiology 58(2):197

Hanowell ST, Kin YD, Jones M et al. (1982) Nitrous oxide addition to halothane and enflurane: depressant or stimulant in coronary artery disease? Anesthesiology 57(3):A77

Hill GE, English JE, Lunn J et al. (1978) Cardiovascular responses to nitrous oxide during light, moderate, and deep halothane anesthesia in man. Anesth Analg (Cleve) 57:84–94

Hornbein TF, Martin WE, Bonica JJ et al. (1969) Nitrous oxide effects on the circulatory and ventilatory responses to halothane. Anesthesiology 31:250–260

Hornbein TF, Eger EI II, Winter PM et al. (1982) The minimum alveolar concentration of nitrous oxide in man. Anesth Analg (Cleve) 61:553–556

James MFM, Manson EDM, Dennett JE (1982) Nitrous oxide analgesia and altitude. Anaesthesia 37:285–288

Kaplan R, Abramowitz MD, Epstein BS (1981) Nitrous oxide and airfilled balloon tipped catheters. Anesthesiology 55(1):71–73

Knill R, Prins L, Strupat J, Clement J (1980) Nitrous oxide and vaporizer outputs: transient or continuous effect? Anesth Analg (Cleve) 59:808–809

Kopp KH, Wehmer H (1981) Tubusverlegung von Silkolatex-Tracheaultuben bei der Verwendung von Lachgas. Anaesthetist 30(11):577–579

Lam AM, Clement JL, Chung DC, Knill RL (1982) Respiratory effects of nitrous oxide during enflurane anesthesia in humans. Anesthesiology 56:298–303

Lampert BA, Gould DB, MacKress TN et al. (1982) The influence of nitrous oxide solubility in volatile anesthetics. Anesthesiology 57:A164

Lappas DG, Buckley MJ, Laver MB et al. (1975) Left ventricular performance and pulmonary circulation following addition of nitrous oxide to morphine during coronary-artery surgery. Anesthesiology 43(1):61–69

Lunn JK, Stanley TH, Eisele J et al. (1979) High dose fentanyl anesthesia for coronary artery surgery: plasma fentanyl concentrations and influence of nitrous oxide on cardiovascular responses. Anesth Analg (Cleve) 58(5):390–395

MacGillivray RS (1982) Pneumocephalus as a complication of posterior fossa surgery in the sitting position. Anaesthesia 37:722

Mallamo, Maj JT, Hubbard RB, Boone MC, Lt. Col, Stephen C et al. (1975) Expansion of an air-filled subdural space during nitrous oxide anesthesia. Radiology 115:369–372

Man A, Sega S, Ezra S (1980) Ear injury caused by elevated intratympanic pressure during general anaesthesia. Acta Anaesthesia Scand 24:224–226

Moens Y, DeMoor A (1981) Diffusion of nitrous oxide into the intestinal lumen of ponies during halothane-nitrous oxide anaesthesia. Am J Vet Res 42(10):1751–1753

Morton M, Duke PC, Ong B (1980) Baroreflex control of heart rate in man awake and during enflurane and enflurane-nitrous oxide anesthesia. Anesthesiology 52:221–223

Munson ES (1971) Effect of nitrous oxide on the pulmonary circulation during venous air embolism. Anesth Anal (Cleve) 50:785–793

Munson ES, Stevens DS, Redfern RE (1980) Endotracheal tube obstruction by nitrous oxide. Anesthesiology 52:275

Murphy MR, Hug CC Jr (1982a) The enflurane sparing effect of morphine, butorphanol, and nalbuphine. Anesthesiology 57:489–492

Murphy MR, Hug CC Jr (1982b) The anesthetic potency of fentanyl in terms of its reduction of enflurane MAC. Anesthesiology 57:485–488

Nunn JF (1969) Applied respiratory physiology, 1st edn. Butterworths, London

Nunn JF, O'Morain C (1982) Nitrous oxide decreases motility of human neutrophils in vitro. Anesthesiology 56:45–48

Orchard CH, Sykes MK (1980) Errors in oxygen concentration analysis: sensitivity of the IMI analyser to nitrous oxide. Anaesthesia 35:1100–1103

Owens WD, Gustave F, Sclaroff A (1978) Anesth Analg (Cleve) 57:283

Perreault L, Rousseau P, Garneau JF et al. (1981) Problème de la diffusion gazeuse dans l'oreille moyenne au cours de l'anesthesia pour tympanoplastie. Can Anaesth Soc J 28(2):136–140

Perreault L, Normandin N, Plamondon L et al. (1982) Tympanic membrane rupture after anesthesia with nitrous oxide. Anesthesiology 57:325

Piernan S, Roizen MF, Severinghaus JW (1979) Oxygen analyzer dangerous – senses nitrous oxide as battery fails. Anesthesiology 50:146

Saidman LJ, Eger EI II (1965) Changes in cerebrospinal fluid pressure during pneumoencephalography under nitrous oxide anaesthesia. Anesthesiology 26:67

Schulte-Sasse U, Hess W, Tarnow J (1982) Pulmonary vascular responses to nitrous oxide in patients with normal and high pulmonary vascular resistance. Anesthesiology 57:9–13

Shapiro HM, Yoachim J, Marshall LF (1982) Anesth Anal (Cleve) 61(3):304–306

Smith NT, Calverley RK, Prys-Roberts C et al. (1978) Impact of nitrous oxide on the circulation during enflurane anesthesia in man. Anesthesiology 48:345–349

Srivastava S (1980) Tympanic membrane rupture during nitrous oxide anaesthesia (letter). Br J Anaesth 52(9):961

Steffey EP, Gillespie JR, Berry JD et al. (1974) Circulatory effects of halothane and halothane-nitrous oxide anesthesia in the dog: controlled ventilation. Am J Vet Res
35(10):1289–1293

Steffey EP, Johnson BH, Eger EI II (1980) Nitrous oxide intensifies the pulmonary arterial
pressure response to venous infection of carbon dioxide in the dog. Anesthesiology
52:52–55

Stoelting RK (1971) The effect of nitrous oxide on halothane output from Fluotec mark
2 vaporizers. Anesthesiology 35:215

Waskell L, Watson JE, Stokstad ELP, Eger EI II, Koblin DD (1982) Nitrous oxide inac-
tivates methionine synthetase in human liver. Anesthesia Analgesia (Cleve) 61:75

Welch WD, Zaccari J (1982) Effect of halothane and nitrous oxide on the oxidative activity
of human neutrophils. Anesthesiology 57:172–176

Measurements of Volatile Gases

G. Rolly and L. Versichelen

When using potent inhalation anaesthetics such as halothane, enflurane or iso-flurane, it is important to have a precise control of the exact anaesthetic concentration inhaled by the patient. The first imperative is, of course, to use a modern and carefully calibrated vaporizer. The Cyprane vaporizers and the Dräger vaporizers are the most frequently used in Western Europe. However, after a certain time deviations and even major deviations from the dial setting are possible and regular maintenance and calibration are mandatory.

It is also necessary to have a good knowledge of the pharmacokinetics of the inhalation anaesthetics, which differ considerably according to the particular anaesthesia system used, open, semi-closed or closed. The complexity of the behaviour of some low-flow or semi-closed systems calls for clinical monitoring of the real anaesthetic concentration inspired by the patient.

Some years ago, a new technique was developed for measuring vapours of different potent inhalation anaesthetics, but with the exclusion of nitrous oxide. This technique is used in the Engström Emma apparatus, which has a transducer that can be introduced, according to wish, into the different parts of the breathing circuit, either at the mouth of the patient or in the inspiratory or expiratory limb of the breathing circuit; it can also be used for direct calibration of the output of the vaporizer. The transducer contains a quartz crystal which is in direct contact with the gas to be measured. The crystal itself is coated with a film that can absorb and release the molecules of anaesthetic gases, whereby its mass increases or decreases respectively, resulting in changes in the crystal's resonance frequency. The change in frequency is proportional to the concentration of the anaesthetic gas. The most convenient way of using the device is to place it at the Y-piece of the anaesthetic system, enabling inspiratory and expiratory concentrations to be measured breath by breath. According to our experience, this apparatus is reliable for clinical measurements of the vapour concentration. However, after a longer-lasting anaesthesia delay in the read-out of decreasing concentrations can be seen, probably due to the physical characteristics of the technique, whereby the anaesthetic molecules are dissolved in the crystals and release takes some time.

The second but more sophisticated technique for measuring anaesthetic vapour concentrations is the mass spectrometer technique, whereby respiratory gases are analysed qualitatively and quantitatively according to their mass number. Besides potent inhalation anaesthetics such as halothane, enflurane, and isoflurane, all other respiratory gases (O_2, N_2, CO_2, argon, and nitrous oxide) are simultaneously and continuously analysed on a breath-to-breath basis. Our experience is restricted to the Centronic quadripole mass spectrometer, namely the anaes-

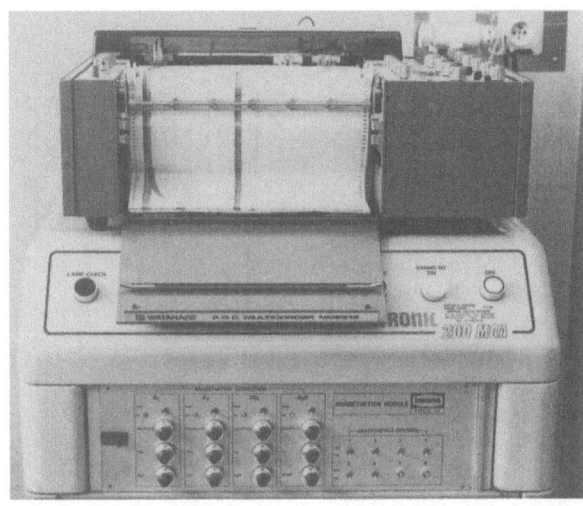

Fig. 1. Centronic mass spectrometer with Watanabe recorder

thetic model with automatic stability control and spectrum overlap eraser for anaesthetic gases (Fig. 1). Halothane is analysed at mass number 116 and enflurane or isoflurane at mass number 51. The respiratory gases are sampled continuously by means of a small-borne catheter positioned ± 5 cm down the endotracheal tube. The sampling rate is of the order of 50 ml/min. The obtained curves can be graphically recorded. The recorder we used allows full-scale height deflection for each gas and also provides a different colour for each gas so that easy recognition is possible. However, for the potent inhalation anaesthetics a different scale going from 0% to 2% is used. Notice also the horizontal time axis and the 1-min marker.

In the first curve obtained in a patient after thiopental induction and ventilation in the open system, a typical wash-in curve for an anaesthetic mixture is shown (Fig. 2). The nitrogen wash-out occurs rapidly after the patient is connected to a 50% O_2/50% N_2O mixture. In a second step, enflurane 1% is introduced into the system, and the actual inspired enflurane concentration is rapidly brought into equilibrium with the concentration delivered by the vaporizer. Due to the tissue uptake, especially during the early induction period, large gradients exist between the inspired and end-tidal concentrations. Similarly, an inspired end-tidal gradient exists for nitrous oxide during the initial uptake.

Commonly, during induction first a high concentration is put on the vaporizer and it is than reduced after some minutes. In this particular situation, first a concentration of 1.8% enflurane was put on the vaporizer and later on reduced to 0.8%. One notices initially the rapid equilibration with the vaporizer concentration but the great difference between inspired and end-tidal concentrations, evidencing the high initial uptake of the volatile anaesthetic; after the change to a lower concentration on the vaporizer, the real inspired concentration quickly equilibrates and the gradient between inspiration and end-tidal concentration now becomes much smaller. And in fact, the end-tidal concentration is only a bit less than in the initial setting.

When switching over to a semi-closed system, as is frequently done in clinical anaesthesia, one immediately sees a rise in nitrous oxide concentration and a de-

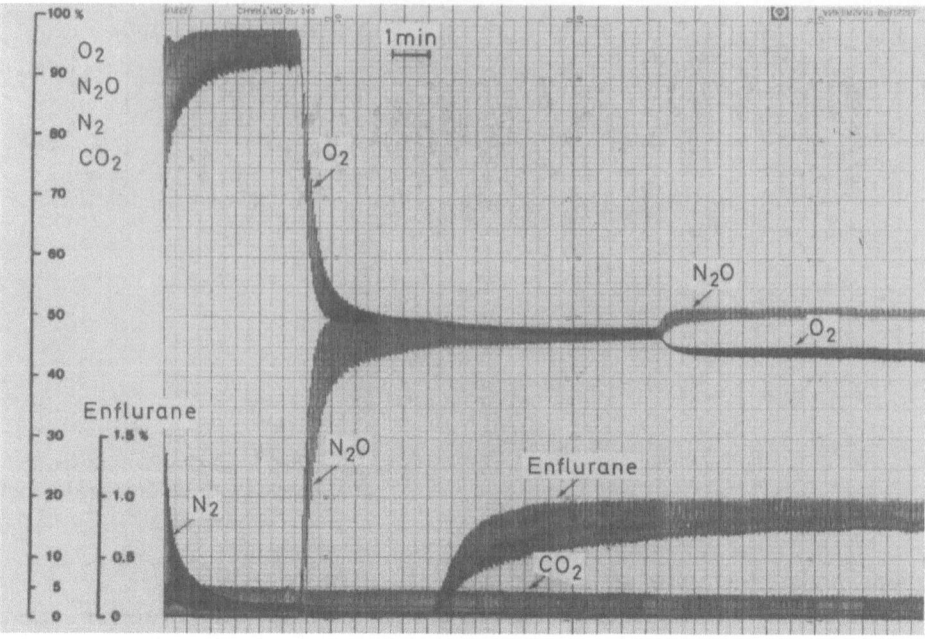

Fig. 2. Anaesthetic uptake in the open system (Rolly and Versichelen 1982). (Current Status of Inhalation Anesthesia, p. 59)

crease in oxygen, although the relative concentrations of nitrous oxide and the inflow of gases are kept rigorously constant. Somewhat to our surprise, little change in the enflurane concentration is noticed.

The pharmacokinetics is different if a semi-closed system is used right from the beginning. Figure 3 shows the induction of a patient ventilated by an Engström respirator without the application of any open-system anaesthetic at the beginning. In this situation a slow elimination of nitrogen from the system and a slow wash-in of nitrous oxide occur. Although the in-flowing gas contains 1% enflurane, the actual inspired concentration is very low at the beginning and induction with the inhalation anaesthetic is prolonged.

The mass spectrometry technique also allows breath-to-breath monitoring of gas exchange, and Fig. 4 shows the breath-to-breath changes between inspired and end-tidal concentrations for the different gases: oxygen, nitrous oxide, CO_2, and enflurane.

A next point of clinical importance is the behaviour of gas concentrations at the end of anaesthesia, and Fig. 5 shows that after first stopping the enflurane administration and then in a second step going over to 100% oxygen ventilation in an open system, both for enflurane and for nitrous oxide, a break point occurs followed by inversion of the gas concentration gradient by the establishment of an end-tidal to inspired concentration gradient.

A very special technique for inducing closed-circuit anaesthesia is the liquid injection of potent inhalation anaesthetics into the circuit. Figure 6 shows that when halothane is injected into the circuit according to the Lowe formula, a seesaw change in the delivered halothane concentration is obtained, together with

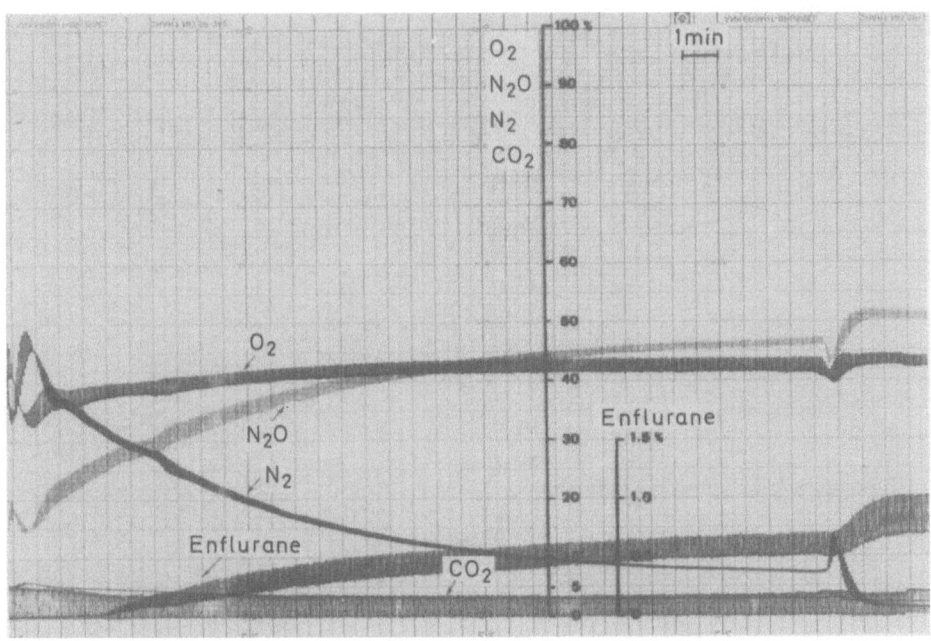

Fig. 3. Anaesthetic uptake in the semi-closed system. (Current Status of Inhalation Anesthesia, p. 60)

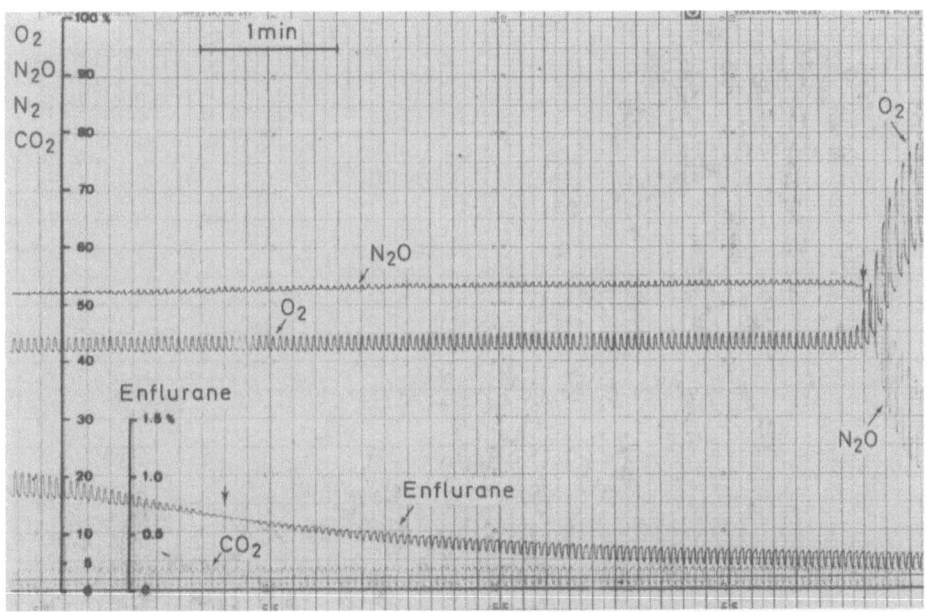

Fig. 4. Breath-to-breath monitoring at the end of anaesthesia (at the *arrows* reversion of the enflurane and N_2O gradients). (Current Status of Inhalation Anesthesia, p. 61)

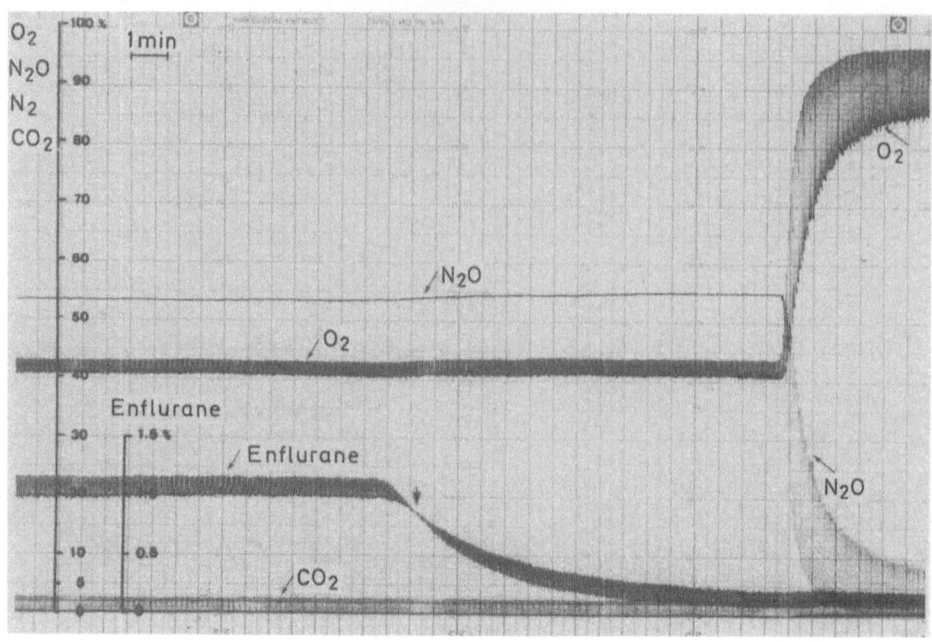

Fig. 5. Anaesthetic wash-out, (at the *arrow* reversion of the enflurane gradient). (Current Status of Inhalation Anesthesia, p. 61)

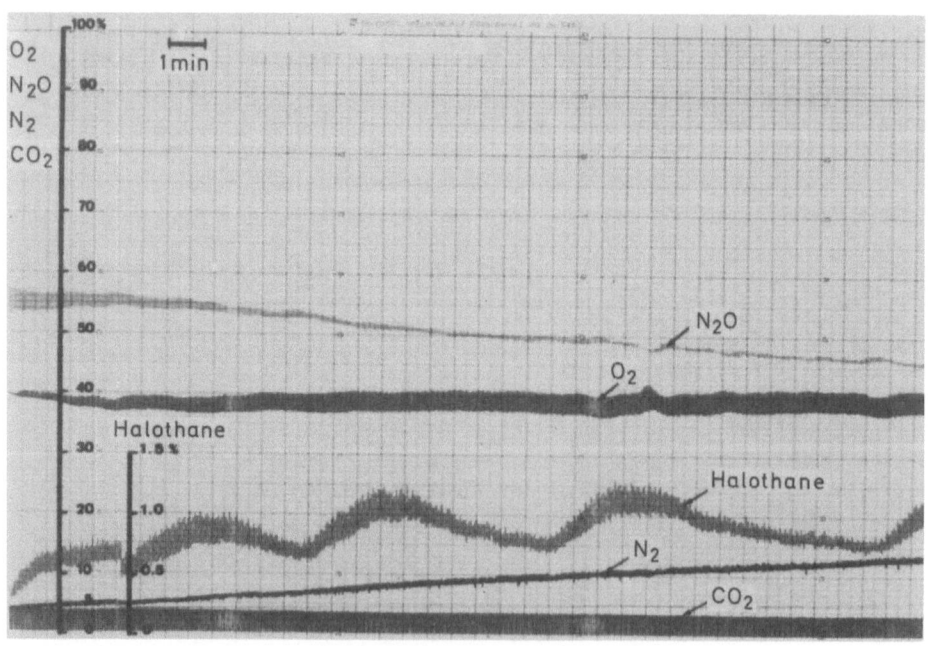

Fig. 6. Closed-circuit anaesthesia by injection of liquid halothane

a progressive rising concentration of nitrogen in the circuit. The same holds true if liquid isoflurane is injected into the circuit together with oxygen 33%/nitrous oxide 66% anaesthesia. The see-saw movements of isoflurane concentrations diminish as the anaesthesia continues, as then the number of injections is decreasing, in accordance with the Lowe formula. These tracings obtained by mass spectrometry help us to gain a clearer understanding in the pharmacokinetics of potent inhalation anaesthetics and of their use in the closed-circuit anaesthesia system.

Reference

Rolly G, Versichelen L (1982) Anaesthetic uptake in the open system. Bunge, Brussels, p 59

The Absolutely Tight Circuit System and the Problem of Excess Humidity

R. Spintge and R. Droh

Two of the most important conditions for closed-circuit anaesthesia are:

1. The absolute tightness of the circuit and the respirator
2. Removal of excess humidity from the system.

These two conditions can be realized as follows. Every patient must be intubated. All connections of the system must exactly fit one another. However, any of the connections can be the reason for a leakage, so they have to be made airtight with a special paste (Oxygenoex). This paste can be sterilized at up to 150 °C.

The single components of the closed-circuit system must not only be sterilized every day but also cleaned out. It is therefore recommended that the paste be renewed after each cleaning. A common source of leakage is the carbon dioxide absorbers. Here it is important first to rebuild the absorber after the cleaning and than to fill it up with sodium hydroxide.

The tightness of the circuit can be proved in the following way:

The circuit must be closed at both ends and then filled up with the gases.
The closed system is placed under water and watched to see if bubbles occur.
The tightness of the system only needs to be proved this way once.

In the daily routine the tightness of the system can be proved by filling it with gas up to a pressure of 40 cmH$_2$O. Once the system is closed, this pressure must be stable for at least 1 min.

A tight system must be handled very carefully. No part of it should be exchanged with parts of other systems. This means that every tight system must always be sterilized as a complete set and afterwards rebuilt at once.

The monitoring of the respiratory gases nowadays is easily done with monitors from Dräger, Datex or Engström. It is also important that all the connections of the sensors with the gas tubes are tight. Here difficulties often arise due to differences in the diameter of connector and tube. Short pieces of silicon rubber tube are used as connectors between sensor and circuit. Of course, the amount of gas which is taken out from the circuit for gas measurements must be replaced. Otherwise, this would be lost and the volume of the gases in the circuit would be reduced.

Besides manual ventilation there should be the possibility of using a respirator. It is possible to make some respirators tight. We were able to do this, for example, with Engström's ER300 and the Dräger Spiromat.

To be used in the closed-circuit system, a ventilator has to fulfill the following conditions:

It must be possible to lower the frequency to below eight breaths per minute.

The rotarmeter must be able to measure in steps as little as 10 ml.

The volumeter must measure exactly with in a range of at least 10 ml.

During changeover from automatic ventilation to manual ventilation the system must remain closed. It must also be possible to use it as a semi-closed system.

The ventilator must have an outward gas bag as an additional gas reservoir so that one can always see whether the system is properly filled with gas or not.

All the connections inside and outside the ventilators must be as absolutely airtight as the anaesthesia machines.

There is no need for a heater and a humidifier, because the patient himself warms up and humidifies the circulating gas.

In order to withdraw excess humidity we use tubes which have an even inner surface. With these tubes water condensers are not necessary. Water-traps are placed at the lowest level of the tubes. One has to be careful to put the water-traps before the sensors measuring O_2, CO_2, and volatile anaesthetics in the circuit, otherwise the measurements will be disturbed. During anaesthesia the water-traps should be emptied when they are half-full.

Fresh gases should enter the circuit before the sodium hydroxide absorbers. The dry, fresh gases collect humidity from the sodium hydroxide and prolong its capacity. The gases themselves are physiologically humidified at the same time. With the increasing humidification of the circulating gases, the sodium hydroxide loses its capacity to absorb the expired CO_2. In this case the sodium hydroxide should not be thrown away, but dried and kept for subsequent use.

To prevent contamination of the circuit by microbes, filters are used in the inspiratory and expiratory parts of the circuit system. These filters can be used for a period of 80 h if sterilized daily. Incidentally, these filters not only protect the system against microbes but also calm the gas stream in the system and mix the gases, so that a reliable measurement of anaesthetic gas concentration is obtained.

Besides all these there are still many more things the industry should improve with respect to the closed circuit. For instance, it should be possible to measure the humidity and temperature of the circulating gases, and new respirators should be more ergonomically designed. The most important task will be to improve the common standard of anaesthetic machines and respirators.

Automatic Ventilation in Minimal Flow Anaesthesia

J. Baum and U. Schneider

Summary

A number of ventilators including Pulmomat 19.1, Spiromat 650, Ventilog, and AV-1 were studied with respect to their use in minimal flow anaesthesia. It was shown that using fresh gas flows of 300 ml O_2 and 200 ml N_2O per minute the new ventilators can be used in minimal flow anaesthesia. Respiratory characteristics and respiratory volumes remain roughly constant.

Due to lekages and a lack of precise rotameters none of the ventilators included in this trial are suitable for closed-circuit anaesthesia.

A Closed Circuit

T. E. J. Healy, P. C. W. Beatty, and B. Kay

The use of expensive anaesthetic agents and the need to reduce operating-theatre pollution has promoted the use of circuits which minimize gas usage. A closed circuit has been developed in the University of Manchester (Beatty et al. 1982). The circuit maintains a constant volume at end expiration. This is achieved by the opposing action of a spring mechanism attached to a bellows of low mass.

The silicone rubber bellows, which has a metal base-plate, is suspended in a bottle which during intermittent positive-pressure ventilation is connected to a ventilator. Attached to the bellows base-plate are two springs. The upper spring mounted inside the bellows is a constant force spring, i.e. a spring whose tension is independent of extension. This is opposed by the action of a spring mounted outside the bellows which has Hooke's law characteristics, i.e. its tension is proportional to extension. Under the influence of these two springs the bellows will rest at a position of equilibrium such that the tension in the Hooke's law spring balances the tension in the constant force spring.

Carbon dioxide absorption reduces the volume of the circuit gases during expiration. As a result the bellows cannot fall back to equilibrium, and the base-plate remains above its equilibrium position. The net force downward on the base-plate, due to the tension in the Hooke's law spring, exceeds that in the constant force spring. The subatmospheric pressure developed in the bellows opens the demand valve and allows nitrous oxide and oxygen in a preset mixture to enter the circuit. The amount of oxygen delivered in the mixture may be adjusted to provide exactly the patient's requirements. The demand valve remains open until the pressure in the bellows returns to near its preinspiration level. The circuit thus maintains a constant volume. If the oxygen concentration in the circuit changes this may be adjusted manually by adding oxygen to the circuit through a low flow rotameter.

The circuit has been shown to work well in clinical use and to be suitable for the study of gas exchange during anaesthesia.

Reference

Beatty P, Wheeler MFS, Kay B, Greer W, Cohen AT, Parkhouse J (1982) A versatile closed circuit. Br J Anaesth 54:689

Automatic Ventilation During Closed-Circuit Anaesthesia

R. M. Schepp, W. Erdmann, B. Westerkamp, and N. S. Faithfull

Introduction

Closed-circuit anaesthesia (CCA) was used as early as 1924, by Waters, using a to-and-fro system. However, lack of proper monitoring facilities and insight into the dynamics of gas uptake and distribution prevented CCA from becoming generally accepted for the safe administration of anaesthesia. The situation has since changed and it is today possible continuously to monitor oxygen, carbon dioxide and anaesthetic vapour concentrations. Additionally, knowledge of pulmonary pathophysiology has increased tremendously, with the result that employment of CCA may now be regarded as just as safe a way of administering anaesthesia as any other.

There are many anaesthetists throughout the world who still routinely employ CCA and who have learned to appreciate the advantages of the CCA technique. These include:

1. Negligible pollution of the operating-room
2. Substantial reduction in the cost of anaesthetic gases and vapours
3. Optimal temperature and humidity of the inspired gases
4. The possibility of easily and efficiently performing continuous monitoring of the uptake of oxygen, nitrous oxide, and anaesthetic vapours.

Nevertheless, some disadvantages do exist and these include:

1. The need for manual ventilation of the paralysed patient. This is due to the fact that, at the moment, no suitable automatic ventilation systems exist that are suitable for use with a completely closed anaesthetic system.
2. Constant adjustment of oxygen inflow is required, usually based on the measurement of oxygen concentration in the inspiratory or expiratory limb of the circuit. Nitrous oxide supply to the circuit is then determined by the volume status of the circuit as determined by the degree of filling of the reservoir balloon.

Unfortunately, for most anaesthetists the disadvantages of CCA have prevailed over the advantages, and consequently this technique has not yet become generally accepted in modern anaesthesia. In the Department of Anaesthesia of the Erasmus University at Rotterdam, a new type of anaesthesia machine has been developed which was primarily designed to overcome any disadvantages, whilst making maximal use of the advantages of CCA. This involved the development of a closed rebreathing circuit with the following design characteristics:

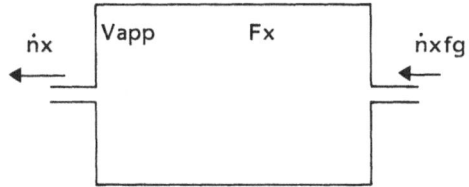

Fig. 1. Gas movement in a closed circuit, where: V app = circuit volume; Fx = fractional concentration of gas; nx = mass uptake of gas x (lost from the circuit or taken up by the patient); $nxfg$ = mass inflow of gas x (fresh gas flow)

1. It should deliver constant inspiratory gas concentrations.
2. A constant end-expiratory circuit volume should be maintained.
3. It should be possible to use the circuit during spontaneous ventilation (with or without continuous positive airway pressure) or during manual ventilation. In addition, it should be possible to use mechanical ventilation in various modes, such as intermittant positive-pressure ventilation or positive end-expiratory pressure.
4. It should be possible accurately to measure gas uptake, for instance of oxygen and nitrous oxide.

Characteristics 1, 2, and 4 are closely related in a closed system.

When considering Fig. 1, it will be apparent that if the mass uptake of gas x ($\dot{n}x$) is equal to the mass inflow of gas x ($\dot{n}xfg$), then the mass of gas x in the circuit will remain constant. In other words:

$$Fx \cdot V \text{ app} = \text{constant.}$$

The fractional concentration of gas x in the circuit multiplied by the circuit volume (V app) remains constant. Conversely, if both V app and Fx are kept constant, the mass of gas x in the circuit remains constant. This implies that $\dot{n}x$ must equal $\dot{n}xfg$. In the case of the two gases x_1 and x_2 (e.g. O_2 and N_2O): if Fx_1 and Fx_2 and V app are kept constant, then $\dot{n}x_1 = \dot{n}x_1fg$ and $\dot{n}x_2 = \dot{n}x_2fg$. As $Fx_2 = 1-Fx_1$ it is sufficient to keep Fx_1 and V app constant in order to keep the mass of both x_1 and x_2 (contained in the closed circuit) at a constant level.

In the classic circle circuit, keeping the mass of a gas contained in the circuit constant presents two difficulties. In the first case, neither the reservoir balloon (in the case of manual ventilation) nor the concertina bag (in the case of mechanical ventilation) are suitable for sufficiently accurate volume measurements to enable feedback control of the circuit volume. Secondly, in the classic closed circuit different fractional gas concentrations are measured in different parts of the system. In addition, during non-steady-state conditions of gas uptake changes in gas concentration at one point in the circuit do not necessarily reflect changes in the mean circuit concentration.

Methods

The Closed Circuit

The water-sealed spirometer, as used in most lung function laboratories, is a rebreathing circuit capable both of accurate volume measurement and of accurate measurement of the mean circuit concentrations of gas mixtures. The latter is achieved by means of a circulation pump (circulating at 70–100 l/min), which ensures very rapid mixing of gases in the rebreathing circuit. Moreover, the employ-

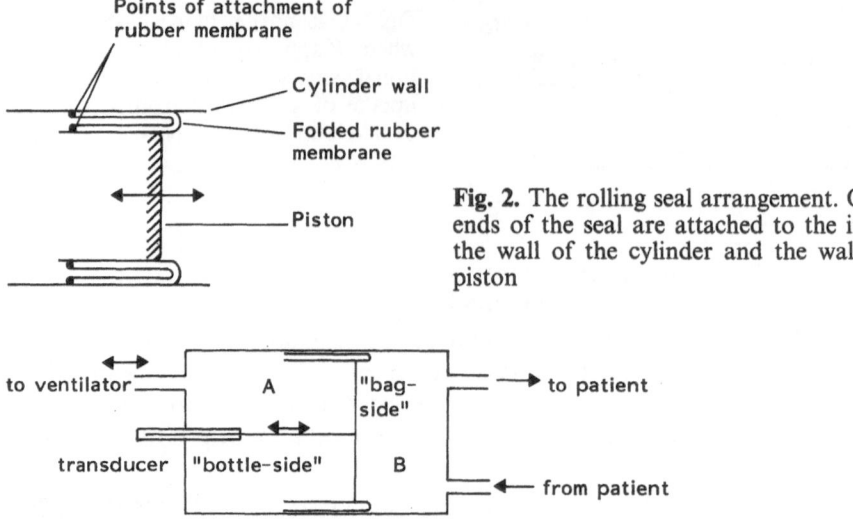

Fig. 2. The rolling seal arrangement. Only the ends of the seal are attached to the inside of the wall of the cylinder and the wall of the piston

Fig. 3. The rolling seal in the "bag in a bottle" arrangement

ment of this circulation pump prevents rebreathing and hence obviates the necessity for the presence of unidirectional valves in the system. This considerably reduces the dynamic resistance of the system.

The water-sealed spirometer is not easily adapted to mechanical ventilation or anaesthesia systems. A rolling seal dry spirometer should, however, prove easily adaptable for these purposes. Basically, it consists of a metal cylinder in which a very light piston is mounted. A thin folded rubber membrane (the "rolling seal") allows for easy movement of the piston within the cylinder. This arrangement is shown in Fig. 2. The dynamic and static properties and the volume accuracy of the rolling seal spirometer are equal to, or better than, those of the water-sealed spirometer.

The rolling seal unit has been employed during anaesthesia in such a way that it acts as a "bag in a bottle" unit, as illustrated in Fig. 3. The complete system is shown in Fig. 4 and incorporates a carbon dioxide absorber, a circulation pump and solenoid valves regulating O_2 and N_2O inflow. The position of the piston is sensed by a linear resistive displacement transducer with high resolution and linearity.

The system functions as follows: compartment A of the rolling seal unit is connected to a mechanical ventilator or, in the case of spontaneous ventilation, to room air. During the inspiratory phase of the ventilator a volume of air is injected into compartment A, which will displace the piston and cause an equal volume of gas to leave compartment B and inflate the patient's lungs. During spontaneous ventilation, a volume of gas is drawn from B causing an equal volume of air to enter A. Due to gas uptake in the lungs, the volume of gas returning to B during expiration will be somewhat less than the amount inspired. As a result, the position of the piston at end expiration will be different from that at the beginning of inspiration. This end-tidal deficit is detected by the transducer and the volume is supplemented with N_2O. At the same time, the present oxygen concentration in the circuit is kept constant via a servo-mechanism operated from the paramagnetic oxygen analyser.

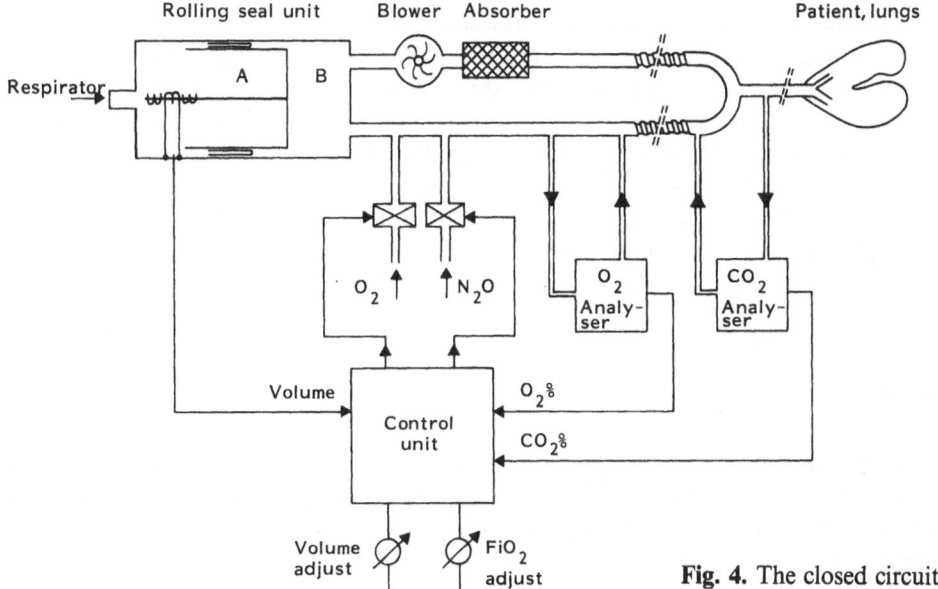

Fig. 4. The closed circuit

The O_2 and N_2O fresh gas flows are regulated by a small digital computer (based on the 8085 CPU). This also calculates and displays the uptake of oxygen ($\dot{V}O_2$) and nitrous oxide ($\dot{V}N_2O$). In addition, the following parameters can be displayed: inspiratory and expiratory minute volumes, breathing rate, inspired oxygen fraction (FIO_2) and airway temperature and pressure. Additional high capacity inlet ports for O_2 and N_2O and a low-resistance gas outlet port can be opened by means of manually controlled solenoid valves. This enables the microprocessor to be overruled at any time.

In cooperation with a company specializing in lung function apparatus (Mijnhardt, Holland), a prototype closed-circuit anaesthesia machine has been constructed. This, after satisfactory laboratory testing, has since been used in the operating room.

Patient Treatment

All patients in the study belonged to American Society of Anesthetists (ASA) groups I or II, and most of them were undergoing plastic surgical procedures. Operating times varied between 30 min and 3 ½ h. All patients received premedication consisting of an oral dose of 1 mg lorazepam the evening prior to surgery and 0.5 mg atropine sulphate intramuscularly 30 min before induction of anaesthesia. Prior to induction, the patients were "denitrogenized" by breathing pure oxygen for 3–5 min.

Induction of anaesthesia was performed using fentanyl 1.5 μg/kg and sodium thiopentone 4 mg/kg. Muscular relaxation was obtained using either pancuronium bromide or vercuronium bromide 0.1 mg/kg. Immediately after intubation the patients were connected to the closed-circuit anaesthesia machine, which had been prefilled with the required O_2/N_2O mixture. Minute volume and frequency were adjusted to obtain an end-tidal carbon dioxide concentration, measured on a Siemens Ultramat M CO_2 capnograph, of between 4.5% and 5.0%. Enflurane

51

was added (in liquid form) to the circulating gas stream in 0.5–1.0 ml doses and its end-tidal concentration was monitored with an Engström "Emma" anaesthetic vapour analyser. At the end of the operation, enflurane could be removed from the gas phase by absorption by activated carbon. Nitrous oxide was removed at the end of the procedure by repeated flushing with oxygen during the expiration phase. Inspiratory and expiratory carbon dioxide were continuously recorded on the writer of the Siemens capnograph, and the FIO_2, circuit volume and $\dot{V}O_2$ were continuously registered on an MFE Industries Series 1000 strip chart recorder. All patients were extubated when awake and breathing spontaneously into the closed circuit.

Results

Throughout the duration of 50 anaesthesias, the preset FIO_2 was maintained at 0.30. However, the actual concentration was observed to oscillate around this value by less than 0.01, the amplitude depending on the individual patient's O_2 uptake. These oscillations were, in fact, the result of the yet imperfect algorithm regulation of the O_2 fresh gas inflow. This same imperfection caused the O_2 fresh gas inflow to be intermittent rather than continuous. In a new version of the computer programme recently developed, O_2 inflow is continuous and FIO_2 oscillations (unter test bench conditions) appear to be less than 0.003. Volume changes of the circuit, at end expiration, were negligible except for very short-lived volume changes immediately following injection of enflurane. Similar changes were to be observed on occasions when surgical colleagues leaned on the patient's thorax! During anaesthesia the temperature in the system rose to $\pm 30\,^{\circ}C$, and the appearance of water droplets in all visible parts of the system indicated a high relative humidity in the circuit.

Discussion

The CCA system described above has proved to function satisfactorily in daily practice. The first aim of development, namely an anaesthetic machine that can deliver anaesthetic gases and vapours in a completely closed system, while the patient is automatically ventilated, has been achieved. The system has proved to be compatible with several modes of ventilation and it has been used with the Siemens Servo 900B ventilator, as well as with the simpler Dräger Oxylog. In principle, any ventilator can be used in combination with this system. The recently developed computer programme offers continuous measurement of oxygen and nitrous oxide uptake, though it must not be forgotten that dissolved nitrogen will leave the different body compartments and will enter the gas in the closed circuit, thus diluting the nitrous oxide. After several hours (depending on body size and composition), the nitrogen will reach levels of 10%–15%.

As both the ventilatory pattern and the end-tidal volume of the gas contained in the circuit can be kept within very narrow limits, the system is very well suited for many non-invasive techniques based on the rebreathing of inert gases of different solubility. For instance, the pulmonary functional residual capacity may be determined by means of helium or argon equilibration. Cardiac output could be determined by means of acetylene or freon-22 uptake measurements, or estimation of ventilation/perfusion distribution patterns may be performed using modification of the multiple inert gas technique (Wagner et al. 1974).

It is our belief that a CCA system, such as here described, can be used as safely, conveniently and efficiently as any open or semi-closed system, whilst at the same time offering many advantages. The extended monitoring facilities, mentioned above, may in the future prove invaluable for high-risk patients who are required to undergo surgery.

References

Wagner PD, Saltzman HA, West JB (1974) Clinical scope and utility of carbon dioxide filtration in inhalation anesthesia. J Appl Physiol 36:588
Waters RM (1924) Measurement of continous distribution of ventilation-perfusion ratios. Curr Res Anaesth 3:68

Discussion I

Droh: May I start with a question to Dr. Healy and Dr. Rolly? Dr. Rolly showed an increase in N_2 during anaesthesia, Dr. Healy did not. How do you explain this?

Healy: I think the reason why the nitrogen doesn't go up must surely be that the circuit system we use is very leak-proof. Having been flushed out at the beginning, the nitrogen did not go up afterwards. What I did indicate in the end was that during an operation there is a rise in nitrogen once the surgeon has made a large incision. If we compare this with our other results in a more theoretical situation without incision, it seems that nitrogen is absorbed from the tissue of the surgical wound. Therefore, we have a rise in the nitrogen concentration in the circuit shortly after incision.

Droh: You think that the nitrogen is absorbed through the wound from the outer atmosphere?

Healy: Yes, up to the time of incision we had nearly no nitrogen in the system and than suddenly nitrogen rose.

Rolly: Well, I am sorry, I don't agree.
1. We did our measurements in a perfectly airtight circuit system.
2. There is a difference whether you connect the patient to the closed-circuit system from the first second or whether you connect him to the circuit 3 or 4 min after induction of anaesthesia. Just at the beginning the wash-out of nitrogen is higher than it is after some time has elapsed.
3. I would like to state that in some of our studies when we changed from open to closed system to measure nitrogen concentration we had a rise in nitrogen in the closed circuit after about 19 min. It is our feeling that the nitrogen comes out of the tissue.

Healy: How soon after the start of anaesthesia did you see this rise in nitrogen?

Rolly: We did not really have to wait for it. We could see it from the beginning.

Healy: When you started your anaesthesia you washed out nitrogen from the system, from the patient's functional residual capacity and from the tissue. You can rapidly wash out nitrogen from the circuit. You can rapidly wash out the nitrogen which is in the lung in the functional residual capacity of the patient if you have a completely airtight system. Then there is only the nitrogen left which is in the tissue of the patient, and that is about 1 l. One litre of nitrogen can come out to contaminate the gases in the circuit. Once you have washed out the nitrogen you have a clean system. I agree that there is an increase in nitrogen concentration later on, but I don't believe that it comes either from the circuit or from the patient's original nitrogen. This nitrogen comes from the atmosphere through the surgical wound. That is a hypothesis. I have no evidence to confirm this. The only thing is that if you have a limited volume of nitrogen in the system and if you have a truly tight system, than you should be able to get rid of it.

Rolly: Well, I cannot agree with all you have said. Of course, these measurements were made in surgical patients during operation and after induction of anaesthesia, so I cannot reject everything you have said. But even after 19 min in the open system changing to closed circuit we saw small rises in nitrogen.

Faithfull: It is possible that there was a bad perfusion before. Sometimes there happens to be an increase in blood perfusion in the tissue and this could cause an increase in nitrogen because this nitrogen was not washed out before. I think this is very simple to prove. You take an airtight chamber containing nitrogen, put an animal into it, make an incision and see what happens.

Droh: I must say we have sometimes made similar observations to those of Dr. Healy, but only with operations in the open abdomen.

Siedlecky: A question to Prof. McIntyre: you said that you use 0.7 MAC with a mixture of oxygen and air. The question is, why don't you use 1.0 MAC or even more? You must add fentanyl. With a MAC of 1.0 or higher this would not be necessary. Is there any special reason why you use two analgesics?

The second question to Prof. McIntyre deals with Oxford tubes and the cuff pressure. Here we have the problem of keeping the pressure in the cuff low. We tried to keep the pressure low by filling the cuff with saline. But we found that saline is not good for filling the cuff. I think this problem should be solved now. Perhaps we should use single-use tubes.

Another question to Dr. Droh: What necessary adaptation do you envisage to enable the use of the AV-1 ventilator in the closed-circuit system? What are your experiences with closed-circuit anaesthesia in neonates and small children?

And one last question to the other speakers: What impact does closed-circuit anaesthesia make upon organization in the operating-theatres? Do you see the possibility of using this system in all operating-theatres or do we have to select young and healthy patients for this system? How much time do you need to observe the monitoring systems in closed-circuit anaesthesia? Do you believe that the use of closed-circuit anaesthesia will save time? At the moment our young anaesthesists sometimes only watch the monitoring system without realizing that the operation is already over.

Droh: First of all, we are dealing with the improvement of the AV-1. I would not hesitate to use the AV-1 or the Spiromat or any other ventilator for closed-circuit anaesthesia. These ventilators must only be totally tight. You must speak to the industry, as we are doing. It is important that the respirators have a bag as a reserve. Dr. Spintge and Dr. Baum pointed this out. You must always be able to see how the system is filled up with gas. I trust this "monitor" more than any highly sophisticated electronic monitoring. Therefore we ventilate manually. You will realize at once if the system has any leakage. But I am sure that you can make every respirator airtight. This is also economically interesting for the industry. There are about 3500 clinics in Germany alone. If you presume that they all have only one operating-theatre and if you think that making the respirator airtight will cost about 5000 German marks, then you see that we are speaking about many millions of marks.

The second question dealt with anaesthesia for children. We have no experience with anaesthesia for the newborn. But I think that anaesthesia for the newborn especially should be performed with the closed-circuit system. Last year in London at the European Congress of Anaesthesia, Dr. Fuentes and his group from Mexico reported very good results with the closed-circuit system in 600 anaesthesias in neonates.

Dr. Foster has developed for newborns a circuit system which could be an ideal closed-circuit system for infants. Kuhn's anaesthesia system for infants needs a minimum of $4 \, l \, O_2$ and $8 \, l \, N_2O$, i.e. $12 \, l$ dry and cold gases.

I cannot understand why you think we should have problems with nitrogen. Until now we have treated more than 50 000 patients with closed-circuit anaesthesia and we have not had any problem with the cuff of the tubes. Sometimes we have tried to perform anaesthesia with halothane and oxygen only, but we were never satisfied with this and we always came back to nitrous oxide. Why should it be dangerous?

I should add that for halothane anaesthesia we rarely wash out nitrogen. We just start as usual and when the system is filled we close it up and keep the nitrogen inside. Only for neuroleptanaesthesia do we wash out nitrogen, to achieve a better effect of nitrous oxide.

McIntyre: First of all I never suggested that it would be dangerous to use nitrous oxide in the closed-circuit system. We have used this technique ourselves.

I also did not suggest that nitrous oxide itself is a dangerous drug. I only wanted to point out that there are certain clinical circumstances in which the use of nitrous oxide is undesirable. This is a very small proportion of the practice in the average general hospital. I only wanted to say that with the new narcotic analgesics which are available now the advantages of nitrous oxide are becoming fewer and fewer. In my paper I presented the current information on nitrous oxide and suggested that we are perhaps in a process of change at the moment.

In answer to the question why we use fentanyl, the advantage of fentanyl is that you don't have to consider the metabolic consequences of long-term use. At the moment we cannot overlook the fact that higher amounts of halogenetic agents can lead to cardiomuscular changes.

I must admit that I did not know what an Oxford tube was. We use disposable endotracheal tubes. We have small monitors for the cuff pressure, the temperature and the contents.

Droh: Dr. Siedlecky did ask whether the closed-circuit system can be used only for young and healthy patients. This is not so. The closed-circuit system is especially the system of choice for critical patients. Think, for example, of a patient with cardiac problems: you get to know very quickly when he is in trouble. It can be seen when the patient cannot take up any oxygen or cannot eliminate any CO_2, so respiratory and cardiac problems represent a very strong indication.

Pilgenröder: Dr. Droh, one of your arguments is the financial aspect. How can I persuade the director of my clinic to buy the equipment for closed-circuit anaesthesia? In what time can you save the costs, let us say with 4000 anaesthesias per year?

Droh: First of all you should find out what amount of O_2, N_2O, and halothane/ethrane you use per year and what it costs. Then you should first start with minimal-flow anaesthesia and see how much gas you can save. What gas flows do you use?

Pilgenröder: Two to four litres.

Droh: We calculated that we are saving about 200 000 German marks per year in comparison with a semi-closed system. Before we used the closed-circuit system we needed 2700 l O_2 per patient. Now we need 120 l per patient.

What equipment do you need? Basically, you don't need new equipment. Today every anaesthesist should have a monitoring device for O_2, CO_2, ECG, and pulse curve. Even if you want to use a capable respirator you will be able to save the costs of the technical changes within a year. And you should also consider that you are safeguarding your health and that of all your staff in the operating-theatres!

Baum: I think at the moment it is impossible to change a ventilator without changing the ventilation characteristics. We still have no design to change a ventilator from a semi-closed to a closed system.

Droh: That's right. We have no such design yet. But, for example, Engström and Dräger are able to change their respirators to closed system if they are interested in it.

Gradinua: Do you mean to say that we have to intubate every patient using closed-circuit systems? Disposable tubes would create high costs in this case.

Droh: Yes, I think you should intubate every patient, because this is the only way of keeping the airway leakless and free.

Gradinua: Do you intubate every patient? Then you always have to use succinyl and to measure sodium, for example?

Droh: You can do that afterwards if you have a very urgent case. You must always intubate a patient and especially when his status is critical.

Structure and Effectiveness of the Dräger Microbe Filter 644 St

C. E. van der Smissen

In recent years numerous articles have been published on the subject of bacterial contamination of anaesthesia apparatus and lung ventilators, possibly endangering the patient. The conclusions can be divided into three groups:

1. A contamination of lung ventilators when common precautionary disinfection and sterilization measures are observed is unlikely.
2. A contamination of lung ventilators may occur, but it is not expected that this will lead to a loading with germs of the air inhaled by the patient.
3. Contamination of lung ventilators and contamination of the inhaled air cannot be excluded if no additional protective measures are taken.

A suitable additional protective measure is application of a bacteria filter in the expirational or in the inspirational air flow. Although the decision as to how the lung ventilator is to be controlled is obviously made by the user, the manufacturer of the apparatus still has to account for all possibilities. This means that bacteria filters for use in the lung ventilator must be provided. For Dräger lung ventilators and anaesthesia apparatus, therefore, the microbe filter 644 St shown in Fig. 1 is delivered. When this filter was being developed the following requirements had to be taken into account:

1. Very high separation capacity for bacteria and viruses
2. Low breathing resistance
3. Simple handling
4. Low cost.

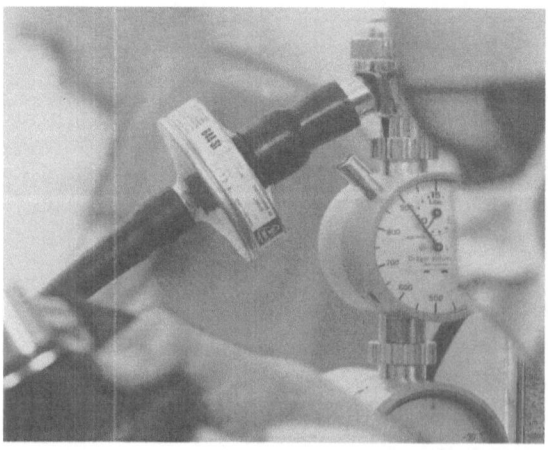

Fig. 1.
Dräger microbe filter 644 St

The set aims were reached as follows. As filter medium a special fibreglass paper with the highest possible capacity is used. It is structured from glass fibres with fibre diameters of from 0.1 to 3 μm and a binder based on silicone resin. The binder gives the paper a high stability and, at the same time, very good water-repellant characteristics. The separation capacity considerably exceeds the value demanded by the DIN 24184 (German standard) for particulate matter filters, class S. Consequently, Dräger received the test certificate from the Dust Research Institute of the main branch of the Industrial Professional Association in Bonn (now known as the Institute of the Professional Association on Occupational Safety), classifying the Dräger microbe filter 644 St as class S.

Low breathing resistance is achieved by utilizing an exceptionally large paper surface of 900 cm^2 (length including both connections 100 mm, diameter 93 mm). Figure 2 shows the connection of the filter. The paper is folded in pleats of 18 mm in height and cut in a circular pleat package. This is then glued into the aluminium casing by means of a specially adjusted polyvinyl chloride (PVC) compound. This structure facilitates a breathing resistance lower than 0.8 mbar measured at 30 l air per minute.

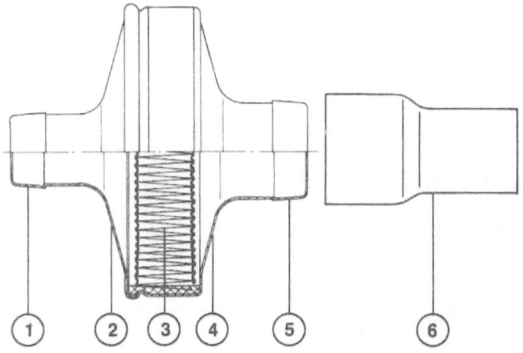

Fig. 2. Cross-section of the Dräger microbe filter 644 St *1*, connection part 24 mm in diameter; *2*, casing lid; *3*, particulate matter filter 92 mm in diameter; *4*, casing; *5*, connection part 31 mm in diameter; *6*, rubber nozzle

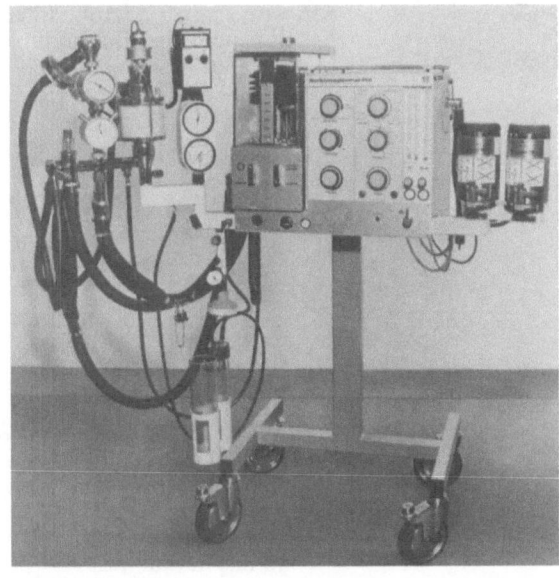

Fig. 3. Dräger lung ventilator 656 with Dräger microbe filter 644 St between expiration tube and expiration valve

59

One connecting part of the filter has a diameter of 24 mm and is intended for connection to the expiration or the inspiration tube. The other connecting part has a diameter of 31 mm and is capped with a rubber nozzle for connection to the lung ventilator. Thus the filter can easily be inserted between expiration or inspiration tube and lung ventilator. Figure 3 shows the lung ventilator 656 with expirationally inserted filter. The filter is fittet between the expiration tube and the expiration valve. Figure 4 again shows the lung ventilator 656, but the microbe-filter is inserted in the inspirational branch. It is fitted between inspiration valve and inspiration tube. Figure 5 shows anaesthesia apparatus Sulla 800 with a filter on the inspiration side. This demonstrates that the filter can be optionally utilized either expirationally or inspirationally. Because of the low breathing resistance filters may also be fitted at both points at the same time.

In order to keep the cost of the use of microbe filters as low as possible, the filter was designed for repeated use. This implies, however, that the filter is sterilizable.

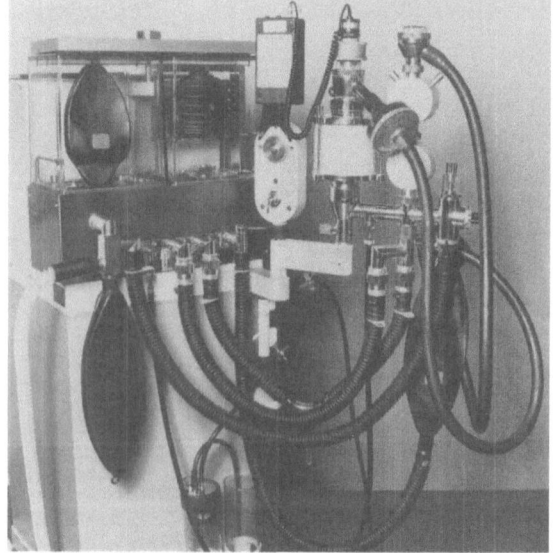

Fig. 4. Dräger lung ventilator 656 with Dräger microbe filter 644 St between inspiration tube and inspiration valve

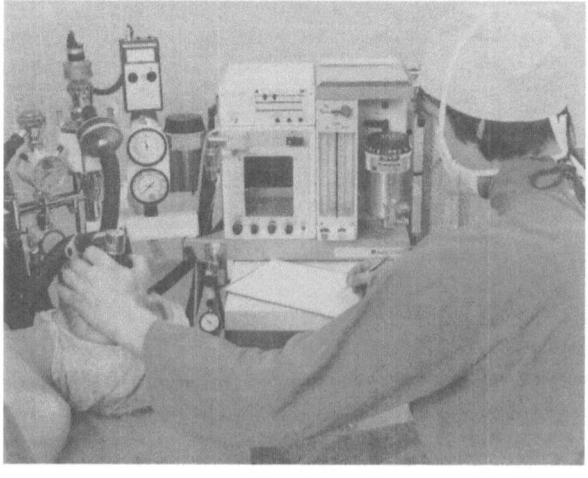

Fig. 5. Dräger anaesthesia apparatus Sulla 800 with Dräger microbe filter 644 St between inspiration valve and inspiration hose

By employing an aluminium casing, fibreglass paper and a PVC casting, a filter was obtained which may be steam-sterilized 20 times at 134 °C. Under standard operating conditions for lung ventilators and anaesthesia apparatus the separation capacity and the breathing resistance do not change significantly during this process. However, it should be taken into account that aerosols of oily therapeutic substances tend to paste up the surface of the filter paper and thus raise the breathing resistance considerably. Microbe filters which have absorbed such substances must not be reused. Also, filters which have been in contact with disinfectant solution must not be reused. Almost all common disinfectant solutions contain substances which cannot be removed by drying. The residue also leads to pasting-up of the paper surface and to an increase in the breathing resistance in the filter. Filters which by mistake have been immersed in disinfectant solution together with the tubes must be rejected.

I should like to refer to the three opinions mentioned at the beginning as to necessity of applying a bacteria filter. In addition to those three to be found in literature, some time ago a fourth opinion came to our attention. Here the subject was the time period between two sterilizations. Commonly, microbe filter 644 St is sterilized after each working day and only in special cases after each application. The opinion expressed stated that this working time is too long. Under conditions of high humidity, as present in a circle system, growth of germs could not be excluded. It was therefore demanded that a bactericide surface be furnished on the filter medium. Experiments were carried out to obtain a microbe filter 644 St equipped with a suitable fibreglass paper. Here we were able to refer to experiences which Dräger has been obtaining for some time with bacteria filters in ventilation systems for operating-theatres and intensive care units. The bacteria filters contain a fibreglass paper which is vapour-plated with copper and silver. The vaporization technique effectively prevents growth of the separated germs. Such a filter medium was now experimentally inserted on the microbe filter 644 St. Particular importance was attached to the fact that sterilization should not destroy the effectiveness of the copper-silver layer. Tests showed that even after 20 sterilization processes the bactericidal effectiveness of the paper surface was fully maintained. Thus today it would be possible to deliver microbe filter 644 St in a design ensuring that between two sterilization processes no germ growth worth mentioning will occur in the filter. Whether such a filter design is necessary and desired by experts is a question I should like to put up for discussion.

In summary it can be said that the microbe filter 644 St is able considerably to reduce the risk of infection in the operation of ventilators and anaesthesia apparatus without entailing too much additional expenditure.

Discussion II

Droh: For many years we have been using Dräger microbe filters with good results. These filters not only keep back microbes, but also help us calm the gas stream in the closed-circuit system, so that we can get reliable measurements of anaesthetic gas concentrations. You said that some people would like to change this filter because they fear that the high gas flow might enable the microbes to migrate through the filter. Do you see this problem in the closed-circuit system, where we have a very low flow? And another question: There are microbe filters which you can use for longer than one day without sterilizing. What do you think of them?

van der Smissen: The coating consists of a very thin coat of copper and silver. The mechanical characterics would not be changed. There is still no evidence that microbes are able to migrate through the filter. In the closed-circuit system you have a high humidity and warmth. I don't know whether this could enable the microbes to migrate through the filter.

Comments on the Environmental Relevance of Commonly Used Inhalation Anaesthetics

G. Müller

Legal Situation

The establishment and operation of installations which, by reason of their nature or their operation, are particularly liable to cause harmful effects on the environment or otherwise to endanger or to cause major disadvantage or considerable nuisance to the general public or the vicinity, shall be subject to licensing.

Act on the Prevention of Harmful Effects on the Environment due to Air Pollution, Noise, Vibration, and Similar Phenomena; short title: Federal Emmission Control Act (BImSchG), Article 4.1.

The Fourth Ordinance for the Implementation of the Federal Emmission Control Act (Ordinance on Installations Subject to Licensing – 4th BImSchV) contains lists of those installations which, under the BImSchG, are subject to a formal or a simplified licensing procedure (58 and 40 types of installation respectively). Hospitals, operating-theatres, and anaesthesia apparatus are *not* included in these lists of the 4th BImSchV. Yet there still is a requirement to comply with the obligations laid down in Article 22 of the BImSchG for the operators of installations not subject to licensing. Under those provisions, even installations not subject to licensing must be established and operated in such a way as to:

Prevent harmful effects on the environment which can be avoided by applying the latest state of technology.
Minimize those harmful effects on the environment which cannot be avoided by applying the latest state of technology.

Decisive importance attaches to the indeterminate legal term "state of technology". As regards medical care, only such facilities and techniques as have been technologically perfected to a degree allowing their use for routine activities and ensuring commercial availability of the apparatus required for them should be regarded as representing the latest "state of technology". In all considerations, the safety of the patient must be given first priority in the selection of the anaesthetic technique.

Situation with Respect to Emissions

The inhalation anaesthetics which are chiefly used in Germany nowadays are halothane (2-bromo-2-chloro-1,1,1-trifluoro-ethane) and enflurane (2-chloro-1,1,2-trifluoroethyl-difluoro-methylether), each combined with nitrous oxide and oxygen.

Halothane and enflurane – halogen compounds administered in gaseous form after evaporation – are contained, in addition to nitrous oxide and oxygen, in the gaseous mixture administered for inhalation by the patient to the amount of about 0.5–2 vol.%.

In the semi-closed respirator system which is in predominant use nowadays, a quantity of 6 l of the gaseous mixture is typically used per minute, i.e. 630 l/h. Assuming an average halothane/enflurane content of 1 vol.%, the hourly consumption of the halogen compound amounts to 3.6 l (gaseous), i.e. some 30 g/h. With a 2:1 ratio of nitrous oxide to oxygen, the consumption of nitrous oxide per hour would amount to 240 l, i.e. 470 g.

By multiplying these figures by the number of tables where operations are performed under anaesthesia, it is possible to compute the mass and volume flows of emitted gases with reference to the hospital or to the operating-theatre.

In that calculation, the used amounts may, by a sufficiently close approximation, be equated with the emitted amounts, since the portions of the anaesthetic which are not exhaled by the patient but are metabolized (chemically degraded) are negligibly low in terms of the issue considered here.

Altogether, some 100 tons of halothane/enflurane are used and emitted in the Federal Republic of Germany every year (halothane/enflurane ratio: about 3:1). The quantity of nitrous oxide used for anaesthesia is about 2500 tons per annum. (These figures were furnished by manufacturers at the request of the Federal Ministry of the Interior.)

Observations on Effects; Comparisons

The acute toxicity of the halogen compounds halothane/enflurane is remarkably low – it is this feature among others which makes them superior to those anaesthetic agents which in the past were generally used and have been replaced by these compounds on the market.

The chronic toxicity cannot, to this day, be explained mechanistically with absolute certainty, but it definitely does exist to a considerable extent. For halothane, the *Senatskommission* (Scientific Committee) set up by the *Deutsche Forschungsgemeinschaft* (DFG) to study work materials detrimental to health has laid down a maximum concentration of 5 ml/m³ or 40 mg/m³ at the place of work – which, even in comparison with other organic halogen compounds, is a very low concentration threshold.

The ecotoxicity of these compounds, i.e. the action of these substances themselves and of their biotic and abiotic degradation products in the environment, does not seem to have been the subject of scientific studies – at least it has not been possible to find any relevant published findings.

The structure of halothane suggests that reduction to trifluoro-acetic acid will take place, while in the case of enflurane there would be reduction to fluoro methoxy-acetic acid:

$$\begin{array}{ccc} F & Cl & F \\ | & | & | \\ F-C-C-Br & \rightarrow & F-C-COOH \\ | & | & | \\ F & H & F \end{array}$$

Halothane has a molecular weight of 197.4 and its boiling point is 50.2 °C.

```
    F   F  Cl              F   F
    |   |  |               |   |
H—C—O—C—C—F  →  H—C—O—C—COOH
    |   |  |               |   |
    F   F  H               F   F
```

Enflurane has a molecular weight of 184.5 and its boiling point is 56.5 °C. The relatively stable degradation products can accumulate in the environment and may produce effects which are not yet known.

The total emitted amount of 100 tons/year appears modest by comparison with the sum of all of the halogenated short-chain hydrocarbons, of which some 250 000 tons/year are used and emitted in Germany (primarily as solvents such as trichloro-ethylene, tetrachloro-ethylene, 1,1,1-trichloro-ethane, dichlor-methane, but also as propellants such as fluorotrichlormethane and other fluor-chlorine hydrocarbons).

With reference to ecotoxicity, mention should be made also of nitrous oxide (N_2O). It has a long life in the troposphere, and in the stratosphere it is transformed into NO, reacting with ozone. Mathematical models have shown that this mechanism accounts most for the reduction of ozone in the stratosphere. At the same time, however, it is assumed that N_2O is produced world-wide on the surface of the earth on a scale of millions of tons per year as a result of microbial processes and thus forms a natural component of the atmospheric cycle.

Conclusions and Outlook

In order to determine the environmental relevance of specific emitted substances, the quantity must be weighed against the ecotoxicity. Substances of which relatively small quantities get into the environment can be very significant ecologically if they are hypertoxic and persistent compounds, e.g. "Seveso poison" – 2,3,7,8-tetrachlorodibenzo-*p*-dioxin (TCDD).

Since the ecotoxicity of the inhalation anaesthetics halothane and enflurane is still largely unclear, a well-founded assessment of the environmental relevance of these substances is not possible at present either.

There is no indication of conspicuous ecotoxicological characteristics suggesting or warranting special administrative measures and specific research efforts. This applies also to nitrous oxide, which according to current knowledge does not cause any considerable additional burden on atmospheric cycles.

However, in view of the application of the general principle of anticipation as part of the environmental policy of the Federal Government, and of the BImSchG provisions on the obligations of operators of installations of any type, it would seem advisable to apply also in anaesthesia the procedure with respect to the lowest emission level and conforming with the latest state of technology. This is undoubtedly the closed-circuit system, which uses and emits less than 10% of the amount of inhalation anaesthetics used in the semi-closed system commonly used at present.

It is to be expected that with the closed-circuit system it will also be possible to provide better protection at the places of work in operating-theatres and to reduce the exposure of the operating staff to anaesthetic gases to a level markedly below that which is tolerated nowadays.

From the point of view of the Ministry responsible for the protection of the environment, the development of the closed-circuit system as a voluntary initiative by the manufacturers and users of such equipment is much to be welcomed.

Discussion III

Droh: Thank you very much, Dr. Müller. We know that it is very difficult to get access to all these figures. In the European Community we use about 700 tons halothane per year. The Americans use about 2000 tons halothane, isoflurane, and ethrane per year. But let as talk about those 700 tons halothane. One millilitre halothane costs at least 0.50 German marks, so 700 tons cost 350 million German marks. Isoflurane is much more expensive still and so is Ethrane.

Each year at least 17 500 tons N_2O are blown up into the atmosphere of Europe, for a price of about 10 German marks per kilogramme N_2O or at least 175 million German marks.

We can say that with the volatile gases and N_2O alone, more than half a milliard German marks per year are blown into the air only to contaminate the atmosphere.

Medical oxygen costs about 7–9 German marks per cubic metre. We don't know how much O_2 is consumed in Europe for medical use but I am sure that this represents another half a milliard German marks per year for Europe. We could thus save about one milliard German marks per year in Europe alone by using the closed circuit with all its advantages.

Two Applications of a Differential Pressure Valve in Anaesthesia

P. A. Foster

It is common practice in electronic design to arrange for a system to compare the values of selected signals and to react should the relationship between these values change. The simplest pneumatic analogy in an anaesthetic breathing circuit is the Heidbrink valve, which dumps anaesthetic gas when the circuit pressure rises above the ambient level. This is a description of two valves in which two values are compared to a third constant value, thus rendering accurate calibration unnecessary and allowing the performance to adapt to changing dynamics.

An Automatic Pollution Control Valve

When a policy decision was made by our hospital authority to institute scavenging of anaesthetic gases in all operating-theatres, an active low-vacuum high-volume displacement system separate from the hospital vacuum system was selected. The various interface devices offered at the time required the anaesthetist to adjust the rate of scavenging flow and/or a controlled leak from the atmosphere into the system, or were extremely bulky in order to contain the largest possible tidal volume that an adult patient might expire.

The design of a small and fully automatic scavenging valve was thus undertaken in our own workshops; the product was coupled with a standard (Penlon) shrouded expiratory valve with 30-mm male taper. This valve is now manufactured in polycarbonate and has been in constant use for about 3 years without need for modification. The basic design is shown in Fig. 1. This is a cylinder with a valve and seat at each end. The valve discs are pressed outwards and held on their respective seats by a common stainless steel spring. In the middle of the cylinder body is a side-arm, to which the scavenging suction line is applied. The valve seats are of different diameters. Thus if suction is applied and both valves are exposed to ambient atmospheric pressure, the disc with the greater area exposed to

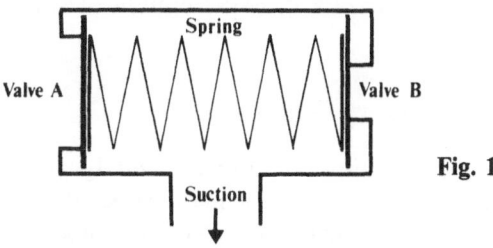

Fig. 1

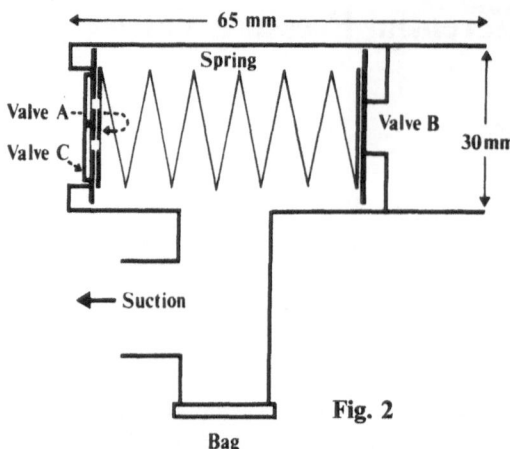

Fig. 2

atmospheric pressure will open. The force applied to this disc is greater, being a product of the pressure differential across the disc and the exposed diameter of the disc. The difference in opening pressures of the two discs is fixed and defined by their different surface areas. A design was chosen in which the opening pressure of the larger-diameter disc was about 1 cmH$_2$O over a large range of flows. The opening pressure differential between the two discs was made to be 0.5 cmH$_2$O.

The practical valve is illustrated in Fig. 2. The larger-diameter disc is exposed to the atmosphere, the smaller-diameter disc is connected directly to the shrouded valve port and scavenging is applied to the side-arm. A small (e.g. 1-l) reservoir bag is also mounted on the side-arm. During use, atmospheric air is continuously entrained by the scavenging flow until breathing circuit pressure rises above atmospheric pressure and opens the expiratory valve. The necessary 0.5-cm pressure differential is thus achieved and expired gas is dumped into the valve. If the volume of expirate is large, or its instantaneous flow is greater than the scavenging flow, the bag temporarily accommodates it. The bag also serves to damp resonances in the system.

When expired gas is dumped, the internal moving components of the valve – the spring and two discs – toggle across to shut off atmospheric air entrainment.

Should there be a failure of scavenging flow, a lightly loaded silicone flap on the larger-diameter disc opens to allow expired gas to pass through the valve to the atmosphere. The performance of this valve has already been described (Moyes et al. 1980, 1981; Foster 1982). It is competent and self-adjusting from scavenging flows up to at least 100 l/min. Optimal scavenging flow is in the region of 25 l/min. The whole assembly is small enough to fit into the palm of one's hand. This scavenging valve is commercially manufactured and obtainable from: Art Medical Equipment (Pty.) Ltd., P. O. Box 23591, Joubert Park, 2044 Republic of South Africa.

A Minimum Percentage Oxygen Delivery Valve

The increasing emphasis on safety factors in life support equipment and the growing realization that operator error, not machine error, is the main cause of accidents, has led to a search for ways of avoiding the delivery of hypoxic gas mixtures

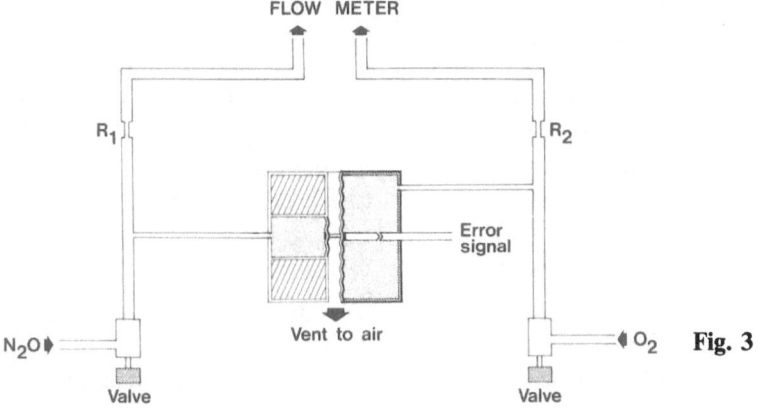

FLOW METER

Error signal

Vent to air

R_1

R_2

N_2O

Valve

O_2

Valve

Fig. 3

to patients. Oxygen monitors are much in fashion, but have a slow response time and have not perhaps yet achieved the desired consistent accuracy and dependability. Also, there are obvious advantages in a purely pneumatic safety device, as opposed to electropneumatic systems, because one depends on only one power source. One such pneumatic device is the gas and oxygen mixer unit which does not permit the delivery of less than a predetermined volume and percentage of oxygen. Its advantage is that it prevents rather than detects operator error.

One such model mixer tested by us, however, seemed to have important disadvantages:

1. It could not deliver gas flows low enough for our routine techniques with circle absorber systems.
2. At its lowest flow rates it showed inaccuracies in excess of 10% of readings.
3. There were doubts as to the long-term reliability of these units, particularly in remote hospitals.
4. Some anaesthetists felt that there were times when it was desirable to override the 30% oxygen lower limit.

An alternative solution was sought in which a minimum of 25% oxygen delivery could be provided down to flow rates of less than 1 l/min, which would be simple and robust in design, and which could be overriden if necessary. Further, the design should be applicable to standard flow meters, thus avoiding the costly rebuilding of machines with the accompanying change in method of operation. These are important factors for consideration in a third world context.

In the first instance the modification was to be implemented on an ergonomically advanced design of machine in which flow meter controls were remote from flow meters. (This allows the machine to be operated in the sitting or standing position by normal staff and by paraplegics in wheelchairs.) The design adopted is illustrated in Fig. 3. Interposed between the flow control knob and the flow meter tube is a laminar resistance unit. Gas source pressure is 4 bar. Between the control knob and the resistance a pressure proportional to flow develops. This pressure can be applied to a control unit.

Two flow-proportional pressure signals, for nitrous oxide and oxygen, are brought to a differential pressure valve and applied to two separate diaphragms with surface areas of ratio 1:3. One side of the diaphragm is exposed to the pressure signal, the other to the atmosphere. The atmospheric surfaces of the two diaphragms are closely approximated and mechanically cross-coupled.

The oxygen pressure signal, being permissibly smaller than the nitrous oxide signal, is coupled to the larger-diameter diaphragm. Should the oxygen to nitrous oxide pressure ratio fall below a figure defined by the mechanical design, the coupled diaphragm assembly moves from its previous static position. An error signal is thus produced under these conditions which can be applied in several ways.

The design selected was one in which:

1. The activating pressure from the oxygen master reducing valve to the nitrous oxide slave valve would be reduced. This was to act as a feedback mechanism by reducing nitrous oxide flow to restore the 25% minimum oxygen concentration.
2. An audible alarm could be sounded. The operator should be aware that the safety system has operated, and thus this audible signal is needed.
3. The error signal could be switched off, allowing the operator to override the safety system.

It should be emphasized that this system was developed independently and possibly at the same time as the Forreger system, in which the error signal is used to augment oxygen flow rather than reduce the nitrous oxide flow.

References

Foster PA (1982) The performance of a scavenging valve. Br J Anaesth 54:359
Moyes DG, Cleaton-Jones P, Foster PA (1980) The design of four anaesthetic scavenging devices. S Afr Med J 58:79
Moyes DG, Cleaton-Jones P, Shaw R, Austin J, Bonner L, Fain H, Mallet J (1981) Comparison of three anaesthetic scavenging devices using cuffed and non-cuffed nasal endotracheal tubes during dental anaesthesia. S Afr Med J 59:180

Measurement and Control of Air Pollution by Anaesthetic Gases and Vapours

A. G. L. Burm, J. Spierdijk, and V. Rejger

Chronic exposure to low concentrations of waste anaesthetic gases is generally regarded as a potential danger to the personnel working in operating-theatres, recovery rooms and other rooms where inhalation anaesthetics are used regularly, as well as to their progeny (Spierdijk and Burm 1982; Cohen 1980). Although the reports on the detrimental effects of waste anaesthetics are not conclusive, it is strongly recommended that the degree of air pollution be reduced as much as possible (Spierdijk and Burm 1982; Cohen 1980; Oehmig 1978; Niosh 1977). Furthermore, it is good practice to check the effectiveness of the systems (scavenging systems or closed circuits) that have been installed to reduce air pollution by measuring the remaining concentrations to which the personnel are exposed. The methods that have been developed in our department to measure exposure levels and to control anaesthetic pollution are described below.

Estimation of the Degree of Exposure to Inhalation Anaesthetics: General Considerations

Several methods have been used to estimate the concentrations of different inhalation anaesthetics to which the personnel are exposed while working in an operating-theatre or recovery room. Most investigators measured concentrations in room air (environmental monitoring). This is a workable approach, but several possible pitfalls should be evaded in order to get reliable results. It has been shown that the concentrations at different locations in one room may differ considerably, due to incomplete mixing of the released anaesthetics with the air in the room (Spierdijk et al. 1975; Burm et al. 1976). In addition it has been demonstrated that the concentration at each location may fluctuate considerably (Burm et al. 1976). In order to get a good estimate of the mean exposure level it is therefore necessary that concentrations be determined in air collected from the breathing zones of the personnel and that samples be collected frequently and at regular time intervals, especially if grab samples are collected (Robinson et al. 1976; Linde and Bruce 1969). In practice this puts a heavy load on the laboratory where the samples are analysed. Therefore, a continuous measurement technique or an integrated personal sampling technique (see below) is to be preferred.

Exposure levels may also be estimated by biological monitoring, for example by measurement of the concentrations of the inhalation anaesthetics in expired air (Linde and Bruce 1969; Hallen et al. 1970) or in blood (Hallen et al. 1970; Gostomzyk et al. 1973) of the personnel. However, in practice this is cumbersome.

Determination of metabolites of the anaesthetics or derivatives thereof in urine samples of the personnel may be a better alternative.

Continuous Montoring of Concentrations of Inhalation Anaesthetics

The problems encountered in the analysis of large numbers of grab samples can be largely avoided if a continuous monitoring technique is used. Knights et al. (1975) have used an electron capture detector to register halothane concentrations continuously. This method is feasible if only one anaesthetic is used and no disturbing compounds contaminate the air. A greater selectivity can be obtained by using a mass spectrometer (Whitcher et al. 1971) or an infra-red spectrometer.

For our investigations a Miran 1A infra-red gas analyser is used. This instrument is provided with a variable wavelength filter (2.5–14.5 μm) and a variable path length gas cell (0.75–20.25 m), enabling the selective measurement of the anaesthetic agents over a wide range of concentrations. The wavelength and path length settings used in our investigations are shown in Table 1.

Continuous registrations are obtained by sucking air from the remote sampling sites (e.g. the breathing zone of the anaesthetist) through the analyser, using the pump, mounted on the analyser, and a length of tubing, provided with a dust filter. The detection limits for both halothane and nitrous oxide are between 0.1 and 0.2 ppm [parts per million: 1 ppm = 0.0001% (vol./vol.)].

Table 1. Wavelength and path length settings of the Miran 1A infra-red gas analyser for the measurement of nitrous oxide and halothane

Anaesthetic agent	Wavelength (μm)	Path length (m)
Nitrous oxide	7.8	0.75
	or 4.5	0.75–14.25
Halothane	12.4	20.25

Personal Sampling Techniques

In the past few years the interest in personal sampling techniques for the estimation of time-weighted averaged (TWA) anaesthetic exposure levels has increased considerably. Several methods for the determination of TWA concentrations of anaesthetics have been described (Niosh 1977; Halliday and Carter 1978; Burm and Spierdijk 1979; O'Sullivan and Houldsworth 1982). These methods include the use of a sampling tube in combination with a sampling pump or a badge.

The method developed in our laboratory (Burm and Spierdijk 1979) includes the use of a small stainless steel sampling tube (length 7.5 cm, internal diameter 4.3 mm) filled with 200 mg Porapak Q. During use in the operating-room or recovery room the sampling tube is placed in a holder, which is attached to the collar of the operating-suit or overall and connected to a sipin SP-2 air-sampling pump by means of a short length of flexible tubing. The pump is carried in a pocket or attached to a belt.

Air is sucked through the sampling tube continuously at a preset rate (5–100 ml/min), depending upon the anticipated duration of the sampling period (varying from several minutes up to 8 h). Organic anaesthetics such as halothane or enflurane which are present in the air are retained by absorption to the Porapak Q. The amounts of the anaesthetics that have been absorbed during the sampling period are determined by means of gas chromatography following thermal desorption (Burm and Spierdijk 1979). The mean concentrations of the anaesthetics in the air are then calculated from the amount of anaesthetic absorbed and the sampled volume. The detection limits for halothane and enflurane are less than 0.1 ppm.

Concentrations in the Air in Operating-Theatres

Concentrations of nitrous oxide and halothane have been measured in 14 operating-theatres. Continuous registration, using a Miran 1A infra-red gas analyser, was applied for the measurement of nitrous oxide and most of the halothane measurements. Concentrations were measured in the breathing zones of the anaesthetists and at several other locations in the room, in general at a height of 1.5 m above the floor. Eleven of the operating-rooms were equipped with a mechanical ventilation (air-conditioning) system. Two operating-rooms were not equipped with an air-conditioning system and one was equipped with an air-cooling system designed for complete recirculation. Anaesthetic agents were mostly administered to the patients via a semi-closed circle system. Most of the patients were intubated and were mechanically ventilated. Typical gas flow rates were 6–10 l/min (O_2:N_2O = 1:2) with 0.5%–1.0% halothane. The personal sampling technique described in the previous section was used incidentally to estimate the levels of exposure to halothane of the anaesthetists.

The results of the investigations are summarized in Table 2. Some of the details have been published elsewhere (Burm et al. 1976). Concentrations were highest in operating-theatres lacking a mechanical room air ventilation or air-conditioning system. The (natural) ventilation rate in these rooms was very low and consequently the concentrations increased steadily during the operation and during consecutive operations. Since the anaesthetics remained in the operating-theatre

Table 2. Typical nitrous oxide (C_{N_2O}) and halothane (C_{hal}) concentrations measured in the breathing zone of the anaesthetist before and after installation of a scavenging system in the operating-theatres

Type of operating-theatre	Scavenging	C_{N_2O} (ppm, vol./vol.)	C_{hal} (ppm, vol./vol.)
Without air-conditioning[a]	No	1500–3000	10 –35
	Yes	100– 300[b]	1 – 3[b]
With air-conditioning	No	200– 500	2 – 5
	Yes	15– 50	0.2– 0.5

[a] Or equipped with an air-cooling system with complete recirculation of the air
[b] Extrapolated values

for a long time, there was ample time for mixing with the room air. This explained the observation that concentrations measured at other locations were very similar to those measured in the breathing zones of the anaesthetists. In operating-theatres equipped with an efficient air-conditioning system a steady-state concentration was reached within 15–30 min. Beyond this time the concentrations fluctuated around a certain mean value, which was different at different sampling sites. In most operating-theatres the highest concentrations were measured in the breathing zone of the anaesthetist, while significantly lower concentrations were measured at more remote locations.

The results of the measurements demonstrated that the concentrations of both nitrous oxide and halothane were diminished drastically after installation of a scavenging system. In our operating-room the levels were reduced to about 10% of the original concentrations.

Control of Air Pollution by Inhalational Anaesthetics in Operating Rooms: General Considerations

The degree of air pollution in operating-rooms can be reduced in several ways, e.g. by using other anaesthetic techniques (regional anaesthesia or total intravenous anaesthesia) whereever applicable, by using closed-circuit anaesthesia or by using a scavenging system. In this paper only scavenging systems will be considered.

In order to be suitable for use in practice, scavenging systems must meet several requirements (Rejger et al. 1981; Burm et al. 1982):

1. They should be simple to operate and not require the attention of the anaesthetist during anaesthesia.
2. They must not interfere with the patient's respiration or with the functioning of the anaesthetic machine and lung ventilator. The build-up of excessive positive or negative pressures at the connection to the anaesthetic apparatus must be avoided (pressure limits -50 to $+200$ Pa).
3. Excess gases collected in the system must be quantitatively transported to the outside air.

The most commonly used scavenging systems use the exhaust grill of the air-conditioning system or the central high vacuum pipeline system or a special ejector (Cohen 1980; Oehmig 1978; Niosh 1977; Enderby et al. 1978; Lund and Osterud 1976; Jørgensen 1974). Passive (through-wall) systems have also been described (Lack 1976; Metha et al. 1977). All these disposal routes have limitations. A passive system may be contra-indicated in windy areas where wind pressures may interfere with the functioning of the system or even with the functioning of the anaesthetic apparatus. Scavenging via the air-conditioning system is worthless if air is recirculated. Finally, the high-vacuum pipeline system may have insufficient capacity and should not be used for scavenging flammable anaesthetics. In many circumstances an independent system, using either an ejector or a small pump for the disposal of the anaesthetics, or an independent central low-vacuum pipeline system may be indicated.

The Leiden Scavenging System[1]

Commercially available scavenging systems are often designed for use with a specific disposal system or a specific anaesthetic machine or lung ventilator and consequently lack versatility. Other systems often do not meet the requirements mentioned in the previous section. Therefore, we developed a multi-purpose scavenging system that can be connected to many of the currently available anaesthetic machines and ventilators and can be used in combination with different disposal routes. During the development period the requirements mentioned above were constantly kept in mind.

A schematic illustration of the Leiden scavenging system is shown in Fig. 1. The interface system shown consists of a metal block with a device for the attachment of a reservoir bag (volume 2 l) and provided with two inlets for connection to the anaesthetic apparatus and/or the lung ventilator and one outlet for connection to the disposal system. Waste anaesthetic gases are fed in the reservoir and subsequently removed from the reservoir by the disposal system. A special outlet connector (Rejger et al. 1981) and normal vacuum tubing are used for connection of the interface to a high-vacuum, low-flow disposal system (central vacuum system, injector or pump). This connector is screwed into the outlet and is constructed in such a way that at a suction rate of 20–30 l/min up to 15 l of waste gas can be removed per minute without leakage of gases into the room. High peak flows are damped by the reservoir and a restriction (diameter 9.3 mm) in the connecter. The difference between the total suction rate and the amount of waste gases supplied is made up by the intake of ambient air via four small holes (diameter 6 mm) in the connector and two holes (diameter 10 mm) in the metal block. The intake of ambient air prevents the transmission of excessive negative pressures. In this way the subatmospheric pressure measured at the connection to the anaesthetic apparatus is limited to -50 Pa (-0.5 cmH$_2$O) at the most, as long as the total suction rate does not exceed 30 l/min. The use of the special connector also prevents the buildup of excessive (positive) back pressures in case of failure of the disposal sys-

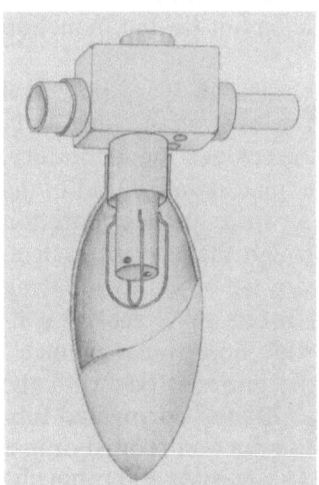

Fig. 1. The Leiden scavenging system: interface, consisting of a metal block with reservoir, two inlets (*upper and left side*) and one outlet (*right side*) for connection to the disposal system. When it is connected to a high-vacuum, low-flow disposal system a surplus of room air is sucked through the two holes (*right bottom*) in the metal block and four holes in the outlet connector. A tight connector is used for connection to a low- vacuum system, e.g. the exhaust grill of air-conditioning is a central low-vacuum pipeline system.

---

[1] Sold by Hoekloos, Schiedam, The Netherlands

tem. Waste gases then escape from the reservoir into the room via the holes in the connector and the metal block. Even in this situation the back pressure measured at the connection to the anaesthetic machine will seldom exceed 200 Pa (2 cmH$_2$O).

The interface can also be connected to a low-pressure disposal system, e.g. the exhaust grill of the air-conditioning system, or be used as part of a passive scavenging system. In this case another outlet connector is used (Rejger et al. 1981), because the intake of room air to prevent the transmission of high negative pressures is not necessary. The connector is then connected to the disposal system using wide-bore tubing to minimize the resistance to gas flow. Both the 19 mm conical inlet connectors and the outlet connectors are connected to the interface by means of screw-thread connections, so that they can be replaced easily. The inlet and outlet connectors are non-interchangeable. The metal block is provided with four screw-holes for the attachment of a clamp, making it possible to mount the system on an anaesthetic machine or ventilator equipped with a rail system. The connection between the scavenging system and the pollution control expiratory valve or the outlet of the anaesthetic apparatus or lung ventilator is made with 19-mm conical connectors and light-weight plastic tubing (internal diameter 20 mm). A 30-mm or 19-mm cone can be used to connect the tubing to the pollution control expiratory valve. It is often possible to connect the scavenging system directly to the ventilator outlet with special links and thus avoid the use of tubing. If only one of the inlets is required, the other can be closed with a blanking plug.

Disposal of Waste Gases Using a Central Low Vacuum Pipeline System

Several years ago, when a new operating-theatre complex was built in our hospital, we decided to install an independent system for the disposal of waste anaesthetic gases and designed a central low-vacuum system. This system is used in combination with the interface described in the previous section (in operating-theatres only) or connected directly to the anaesthetic circuit (in the induction rooms).

The collected waste gases are transported into the open air via a polyvinyl chloride pipeline system (Fig. 2) with terminal points in each of the operating-rooms and in addition in all induction rooms and in the engineering laboratory, making up a total of 27 terminal points. One single fan, placed on the roof of the building, sucks the waste gases and an excess of room air through the connection boxes (Fig. 3) installed at the terminal points and through the pipeline system. The front cover of each connection box is provided with ten holes, each having a diameter of 7.5 mm, through which the excess of room air is sucked continuously. The cover has a provision to connect a wide bore (inner diameter 16.5 mm) male connector, which in turn connects the terminal box with the above-mentioned interface via light-weight wide-bore (20 mm) corrugated tubing. The terminal boxes are installed either in the wall (induction rooms) or in a ceiling pendant (operating-rooms). The total flow rate (waste gases and room air) sucked at each terminal point is 100 l/min. Adjustment of the flow rates at all terminal points was facilitated by installation of a wide-bore main channel with a

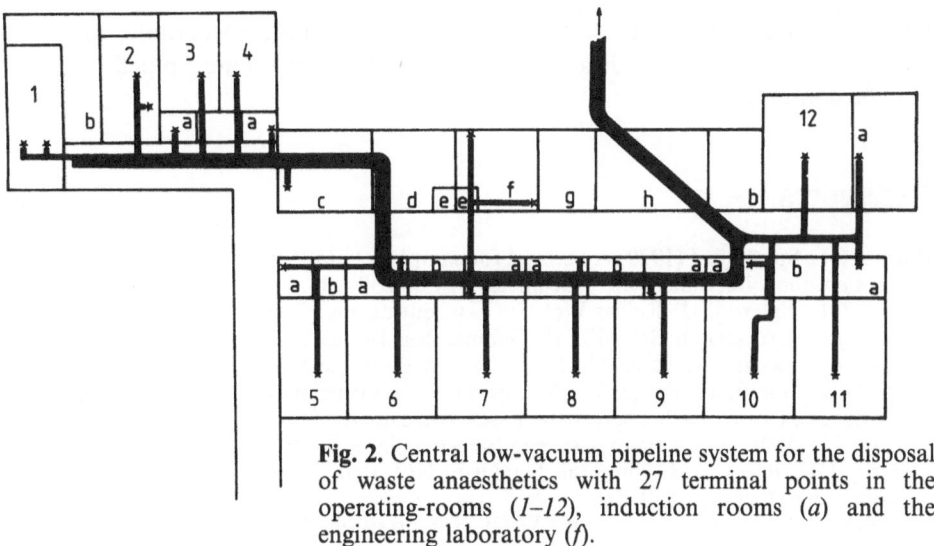

Fig. 2. Central low-vacuum pipeline system for the disposal of waste anaesthetics with 27 terminal points in the operating-rooms (*1–12*), induction rooms (*a*) and the engineering laboratory (*f*).

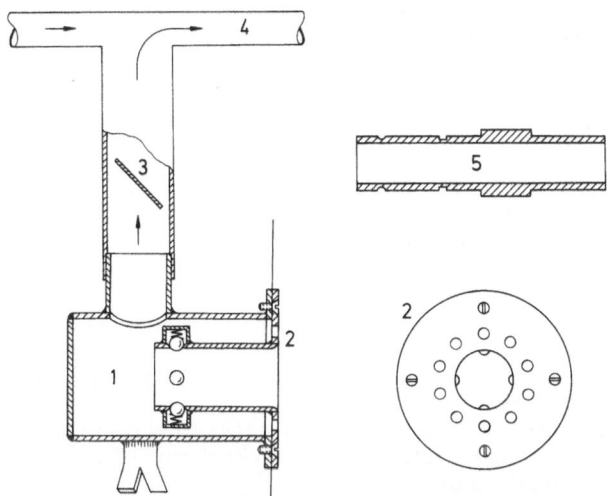

Fig. 3. Connection box, installed at the terminal points, of the low-vacuum disposal system. The front cover (*2*) is provided with ten holes through which room air is sucked and a female connector. A throttle valve (*3*) is installed between the connection box (*1*) and the main channel (*4*). The interface installed in the anaesthetic apparatus is connected to the terminal box using a flexible tube and a male connector (*5*)

small resistance to gas flow, and by balancing the resistances of the tubing and the piping connecting the terminal boxes with the main channel. The flow rates could then be adjusted to 100 l/min by throttle valves, installed between the terminal boxes and the main channel.

The system was designed in such a way that both subatmospheric and positive back pressures were minimized. In practice the subatmospheric pressure does not exceed -11 Pa (0.11 cmH$_2$O), while positive back pressures seldom exceed 50 Pa (0.5 cmH$_2$O).

We conclude that an independent system is the ideal system to scavenge anaesthetics in new operating-theatre complexes or those to be renovated. In existing operating-rooms the costs of the installation may be considerable, and here scavenging via the air-conditioning system or the high-vacuum pipeline system may be considered.

References

Burm AGL, Spierdijk J (1979) A method for sampling halothane and enflurane present in trace amounts in ambient air. Anesthesiology 50:230–233

Burm AGL, Spierdijk J, Rejger V (1976) Concentrations of anaesthetic agents in the air in operating rooms. In: Spierdijk H, Feldman SA, Mattie H (eds) Anaesthesia and pharmacology. Boerhaave Series No 12. University Press, Leiden, pp 263–274

Burm AGL, Spierdijk J, Rejger V, Strokap H (1982) A central low vacuum pipeline system for the disposal of waste anaesthetic gases. Anaesthetist 31:200–203

Cohen EN (1980) Anesthetic exposure in the workplace. MTP Press (Ltd), Lancaster

Enderby DH, Booth AM, Churchill-Davidson HC (1978) Removal of anaesthetic waste gases. Anaesthesia 33:820–826

Gostomzyk JG, Eisele G, Ahnefeld FW (1973) Chronische Narkosebelastung des Anaesthesiepersonals im Operationssaal. Anaesthetist 22:469–474

Hallen B, Ehrner-Samuel H, Thomason M (1970) Measurement of halothane in the atmosphere of an operating room and in expired air and blood of the personnel during routine anaesthetic work. Acta Anaesthesiol Scand 14:17–27

Halliday NM, Carter KB (1978) A chemical adsorption system for the sampling of gaseous organic pollutants in operating room atmospheres. Br J Anaesth 50:1013–1018

Jørgensen S (1974) The injector flowmeter and its clinical evaluation. Acta Anaesthesiol Scand 18:29–33

Knights KM, Strunin JM, Strunin L (1975) Measurements of low concentrations of halothane in the atmosphere using a portable detector. Lancet 1:727–728

Lack JA (1976) Theatre pollution control. Anaesthesia 31:259–262

Linde HW, Bruce DL (1969) Occupational exposure of anesthetists to halothane, nitrous oxide, and radiation. Anesthesiology 30:363–368

Lund I, Østerud A (1976) Ausrüstung zum Absaugen von Expirationsluft während der Narkose. Anaesthetist 25:541–542

Mehta S, Behr G, Chari J, Kenyon D (1977) A passive method of disposal of expired anaesthetic gases. Br J Anaesth 49:589–593

NIOSH (National Institute for Occupational Safety and Health) (1977) Occupational exposure to waste anesthetic gases and vapors. DHEW (NIOSH) Publication No 77–140. US Government Printing Office, Washington DC

Oehmig H (1978) Methoden der Elimination von Narkosegasen und -dämpfen. In: Kirchner E (Hrsg) 20 Jahre Fluothane, Anaesthesiologie und Intensivmedizin, Bd 109. Springer, Berlin Heidelberg New York, S. 158–165

O'Sullivan J, Houldsworth HB (1982) Method for the determination of personal nitrous oxide exposure levels. In: Boulton TB, Atkinson RS et al. (eds) Abstracts of the 6th European congress of anaesthesiology. Academic, London, pp 434–435

Rejger V, Burm AGL, Spierdijk J, Strokap H (1981) The Leiden scavenging system: a simple and versatile apparatus to control air pollution in operating rooms. Br J Anaesth 53:1359–1363

Robinson JS, Thompson JM, Baratt RS, Belcher R, Stephen WI (1976) Pertinence and precision in pollution measurements. Br J Anaesth 48:167–177

Spierdijk J, Burm AGL (1982) Environmental hazards in the operating theatre associated with small amounts of anaesthetic agents. In: Scurr C, Feldman S (eds) Scientific foundations of anaesthesia, 3rd ed. William Heinemann, London, pp 631–636

Spierdijk J, Burm AGL, Bossers PA, van Beukering FC, van Gunst E (1975) Distribution of anesthetic gases in an operating room. In: Henschel WF, Lehmann CH (eds) Schädigungen des Anaesthesie-Personals durch Narkose-Gase und -Dämpfe. Anaesthesiologie und Wiederbelebung, Bd 89. Springer, Berlin Heidelberg New York, S. 44–53

Whitcher C, Cohen EN, Trudell JR (1971) Chronic exposure to anesthetic gases in the operating room. Anesthesiology 35:348–353

Measurement of Pulmonary Capillary Perfusion in Intubated Patients with a Non-invasive N₂O-Rebreathing Method

T. Stokke, H. Boch-Fiola, I. Hänsel, and H. Burchardi

Summary

The study describes a method for the measurement of pulmonary capillary perfusion in intubated animals.

The results were compared with measurements taken with a Swan-Ganz catheter using Fick's formula. The statistical analyses show a difference of $\pm 10\%$ between the two methods.

First clinical trials in intensive care patients show the reproducibility and accuracy of this method to be satisfactory. It must be remembered that this method measures only the perfusion of those parts of the lung which take part in the gas exchange. The perfusion measured by the N₂O-rebreathing method can only be taken as cardiac tidal volume if the gas exchange is normal. This makes it possible to recognize intrapulmonary changes in microcirculation.

Discussion IV

Droh: Dr. Stokke, if I understand you correctly, you can estimate the cardiac stroke volume through the wash-out of N_2O. With the closed-circuit system we could go one step further. Here we know how much oxygen is taken up and with Fick's formula we can calculate the cardiac minute volume (cmv).

$$CMV = \frac{O_2 \text{ consumption (ml/min)} \times 100}{avDO_2 \text{ (vol. \%)}}$$

$avDO_2$ = arterio-venous O_2 content difference.
The wash-out of N_2O gives us the CMV directly, so that we can get the $avDO_2$ and we also have information about diffusion and perfusion etc. Our dream is to get a new parameter through the closed-circuit system.

We can say how much N_2O and how much O_2 have been taken up during a certain time. It should be possible to find a mathematical formula describing cardiac stroke volume beat by beat. At the moment this is only an idea, but it should be possible to realize it in the near future.

Stokke: I think you are a little bit too optimistic. The method we used in Göttingen is a typical non-steady-state method. At the beginning we say N_2O in the blood is zero. If we want to use a similar method during anaesthesia, we must know the real partial gas pressures in the blood. We must know the $avDN_2O$.

But perhaps there is another method which should be propagated: if you apply sinus-shape halothane concentrations inspirationally and if you register this, then you can estimate the $\dot{V}_{A/Q}$ relationship from this curve. The only problem is how to measure real alveolar ventilation ($\dot{V}_A$) You need this alveolar ventilation to calculate the perfusion $\dot{Q}$ with Bohr's formula. Using this method you need monitoring for end-expiratory gas concentration in your anaesthesia system.

Droh: We can measure halothane or isoflurane or Ethrane end-expiratorily.

Schepp: You need still a very complicated apparatus for your measurements of end-expirator halothane concentration. Also the frequency of halothane application is still such that you can only use it with young and healthy patients. If you choose the right frequency you prevent the build-up of back-pressure in the femoral artery. But first you have to find out the right frequency for every patient. This is a little bit complicated. We are working on this problem at the moment.

Droh: Do you believe that it is easier to use halothane because you will come to the steady state later?

Stokke: As far as I know nobody has yet tried with halothane using this non-steady-state rebreathing method. From the theoretical point of view I fear that halothane is too easily soluble in blood. You should try.

Droh: Yes, and we should also try to compare the two methods. Perhaps you can get with one method what you cannot get with the other one.

Stokke: The solubility of N_2O is almost too poor to allow good results. It would be better to use acetylene, for example, but this gas is explosive.

Droh: It is fascinating: with the wash-out curve of N_2O and halothane together with the measurement of O_2 consumption it is possible to measure cardiac stroke volume during the entire course of anesthesia. I think we should work upon this problem with great effort.

A Standardized Anaesthesia Machine

P. A. Foster

A painting of an operation for hysterectomy hangs in the Medical School of the University of Cape Town. The artist was the wife of the district surgeon of one of the major South African cities. It was painted in 1939, shortly after the new Medical School was inaugurated. It was an illustration of the "state of the arts" of that time, in which the anaesthetic machine was an oxygen cylinder and a crude bubble bottle for ether. The mixture was delivered by a latex tube beneath a Schimmelbusch mask.

The technique differed little from that portrayed in extant pictures of turn of the century operating-theatres. The compressed gas and the bubble vaporizer were already in use. If one were to compare this with the first South African anaesthetic, given in 1847, the major difference would be the absence of compressed gas.

Forty years on from 1939, the picture has radically changed. There are multigas rotameters, colour codes, calibrated temperature-compensated vaporizers, ventilators, monitors. One may dare more and achieve more, but the realization has come that this has become a potentially dangerous machine and that the major danger lies in operator error.

There are several reasons for operator error, including inadequate training, fatigue and illness, and wrong judgement. But there is one major cause to which too little attention seems to have been paid either by manufacturers or the developing standards organizations. More consideration must be devoted to ergonomic design and to standardization of layout, as is done in motor car or aircraft design. Many anaesthesia machines appear to have been designed by engineers, who do not spend their lives giving anaesthetics.

Obvious examples of design errors have been that the Boyle's machine was sold for decades in a layout suitable only for left-handed operators; that the machine had to be operated in the standing position, and when one was standing in front of the machine the storage drawers could not be opened.

Awareness of design shortcomings, together with the burgeoning amount of extra equipment that was being routinely fitted to machines, led to the conclusion that some standardization was becoming urgently needed, particularly in a situation where a variety of doctors, many not specialist anaesthetists, provide the anaesthetic service.

In South Africa this led in 1981 to a national symposium representative of all areas of anaesthetic practice to discuss some form of standard design. Remarkably, consensus was reached in one day. Guide-lines were set and made available to any manufacturer. Whilst many of the guide-lines referred to minimum safety requirements, considerable attention was paid to a standardized layout, and it is the purpose of this article to discuss some of the concepts put forward.

There are certain basic premises:

1. South Africa is a third-world country, despite the sophistication of its teaching hospitals. Therefore, any universal design must be simple but capable of upgrading for advanced techniques.

2. Many South African hospitals serve rural populations in areas where road access is not always easy. Servicing thus becomes a problem, and this led to the idea of a compact exchangeable module containing the parts needing regular service. This module should be easily transported and exchanged, and should be applicable to a variety of situations – whether in a ceiling- or wall-hung unit, a portable (e.g. military) unit, a simple floor unit or as the heart of an "anaesthetic work centre" on a larger floor trolley.

3. Standard machine design should be suited to the sitting or standing operator – and thus to the paraplegic – and to the right-handed operator.

4. Basic to the whole standardization concept is a standard "display panel" layout on which the essential functions of the anaesthetic machine were to be represented, or better, on which information concerning the anaesthetic mixture only should be displayed. The prime function of an anaesthetic machine is still to provide an anaesthetic mixture, so this display panel, presenting as it does the most important information concerning the anaesthetic mixture, must be so positioned that it is at all times clearly visible over a wide arc.

5. If the machine is for the right-handed operator, then it becomes inevitable that the display panel must:
   a) Be mounted above any work-surface
   b) Be to the right-hand side of the operator who faces the anaesthetic trolley
   c) Be mounted above controls so that the hand operating the controls does not hide the displays. This can allow all controls of the anaesthetic mixture to be grouped together in a second "control panel", where they are all visible as a group within a hand's span. Certain sequences for siting controls can be worked out, but the prime consideration would be that first adjustment controls be above second adjustment controls. This is shown in Fig. 1, where gas controls are above vaporizer controls (vaporizers do not work without gas flow).

6. The right-handed layout seemed to dictate a functional flow sequence of from right to left for gases in – gases out to patient, with feedback from the circuit to the left (see Fig. 2).

7. Other bulky components which might obstruct the circuit display panel, such as breathing circuits and ventilators, must be mounted to the left-hand side of the machine, where the anaesthetic mixture outflow point should be located.

8. Certain dimensions should be set for work surface, sizes, and heights, control knob diameters and interspacing etc. These are available from various sources, and were possibly first formalized by Le Corbusier, who proposed that man be the module around which machines and buildings are constructed:
   a) About 40 cm for a seat height

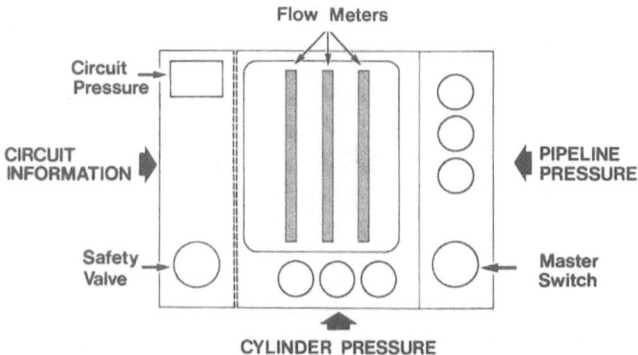

Fig. 1. Proposed formal layout of a display panel on which the relevant information concerning the anaesthetic mixture is collected

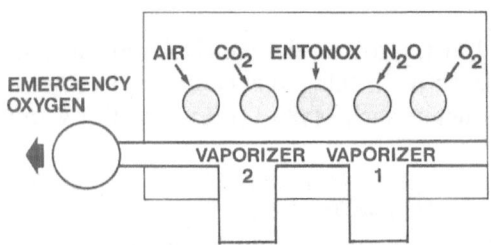

Fig. 2. Proposed formal layout for an anaesthetic mixture control panel which will include emergency oxygen. Ventilator controls can be similarly grouped in an adjacent panel

  b) 70 cm for a work surface above the seat
  c) 90–100 cm for a work-surface for the standing operator with flexed elbows
  d) About 150 cm for a monitor rack
  e) About 220 cm for pendants.
 9. One must consider whether a preset control is a display or a control. It was considered that controls such as a master switch for the machine, or the overpressure relief valve of the circuit, are preset before use and are thus displays.
10. A master switch is one that can inactivate an anaesthetic machine save for oxygen delivery for resuscitation. This should be provided to control the use of potentially dangerous equipment.

Design of a Display Panel

This panel is to be mounted above any work-surface. The most important displays on this panel are thus to be well clear of any work surface, and lead to the layout shown in Fig. 1 using a right-to-left sequence. Primary gas input is normally from pipelines and thus to the right. The master switch would also be on the right.

    The flow meter panel is in the middle and feedback information from the circuit, if provided, would be to the left. The most important feedback from the cir-

83

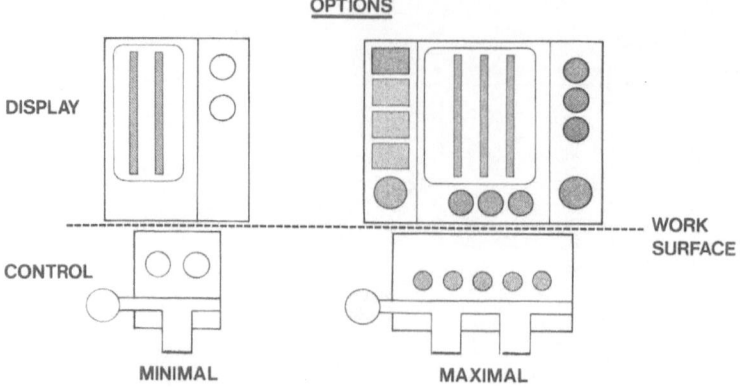

OPTIONS

DISPLAY

WORK
SURFACE

CONTROL

MINIMAL

MAXIMAL

Fig. 3. Illustration of how a display panel may look on a basic machine and a maximum option machine. In the former the optional panel for circuit feedback information is omitted

cuit would be pressure, to be displayed at the top of the panel. An alarm unit might be related to it and situated beneath it. In this panel could be displayed, in the more sophisticated machine, parameters such as pO_2, pCO_2, humidity and temperature. These are all likely to be digital voltmeters of the $3\frac{1}{2}$ digit pattern, for which standard windows can be cut. The preset control of the excess pressure relief valve would appear at the bottom of this panel.

In the simplest of machines, using only two gases and two flow meters, this feedback panel may be omitted (see Fig. 3). The back-up cylinder pressure gauges are mounted just above the work surface and beneath the relevant flow meter in a sequence reading from oxygen on the right, with nitrous oxide immediately left of the oxygen.

Gas Sequences

It was accepted that the oxygen flow meter should be on the right of any unit. Where a multi-gas flow tube unit is used, the left tube should be for an oxygen-carrying gas, e.g. air or Entonox (nitrous oxide). Nitrous oxide should be adjacent to and to the left of oxygen. Just as for flow tubes, a similar sequence must be agreed on for control knobs and gauges, with oxygen to the right or uppermost where vertical sequences are used.

The flow meter, pressure gauge or flow control sequence agreed to, reading from left to right, was:

Air – CO_2/O_2 – Entonox – N_2O – oxygen.

For vertical sequences read:

air = lowest; oxygen = uppermost.

Note: no merit but some danger was seen in supplying pure carbon dioxide on anaesthetic machines. A $50:50$ mixture of CO_2 and O_2 was proposed as "anaesthetic carbogen" as with Entonox (50% N_2 and 50% O_2). The only asphyxiant gas then remaining is nitrous oxide.

84

Design of a Module

For ease of servicing and exchange is was considered that:

1. The minimum module should consist of a suitable standard-size enclosure that can accommodate from one to five flow meter tubes and related gauges.
2. Pipeline connections should be integral with the back of the module using NIST connectors.
3. Reducing valves and oxygen emergency valve should be contained within.
4. The needle valves of the flow meter tubes should be within the module together with any N_2O/O_2-mixture-limiting device that may be optionally fitted. Control knobs might remain on the anaesthetic trolley and interface by some clutch mechanism.
5. Vaporizers, in their present bulky form, should be separate, modular exchangeable items.
6. The width of a module should be such that it subtends a horizontal angle of approximately $15°$ to the view of the operator from the normal working position at a viewing distance of 1 m.

Discussion

The purpose of these design principles is to define a rough standard layout for any anaesthesia machine and to allocate defined spaces on display and control panels of such a machine. The purpose is to present to the user a common design on which training models can be based and emergency drills practised, and which all allow the user who works in different hospitals to move between various makes and models with minimal chance of operator error. It is a specific requirement that the layout should be simple and apposite for simple equipment but that it may be "upgraded" to serve more complex functions.

The complexity of any such "advanced" machines will be in the trolley design, which could, for instance, include ventilators, computer interfaces, a variety of monitors, intravenous infusion pumps and any other complicating machinery that may seem desirable. The only requirement of the mounting trolley would be to provide an adequate working-surface for the anaesthetic tasks, and mounting of all ancillary equipment so as not to obstruct the display and control panels. On the basis of the present trends in anaesthetic technique and monitoring, one may propose the separation of a standard anaesthesia trolley into four or five zones. From above downwards there would be:

1. Monitor displays other than circuit parameters (150 cm upwards)
2. Circuit parameter displays – i.e. display panel (100–150 cm)
3. Control zone (80–100 cm)
4. Storage zone (below 70 cm)
5. A fifth possible "breathing circuit zone" might be considered as a vertical zone on the extreme left of the anaesthesia trolley in which, one below the other, ventilator bellows, circle absorber and suction bottle – the lowest – might be arranged.

This is shown in Fig. 4.

COMPLETE MACHINE

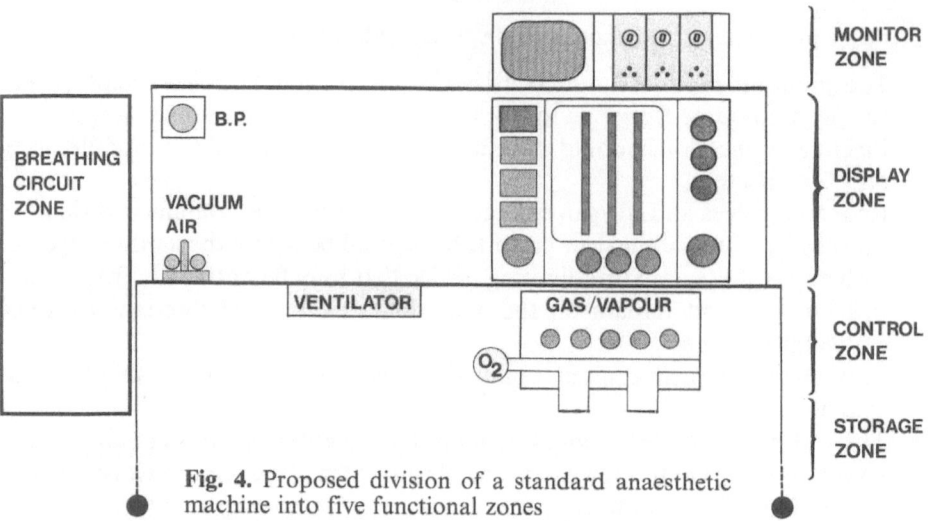

Fig. 4. Proposed division of a standard anaesthetic machine into five functional zones

Anaesthesia machine design will thus follow the example set by the manufacturers of motor cars, aeroplanes, cassette recorders, cameras, and much other commonly used equipment.

Ergonomics and Anaesthesia: Auditory Alarm Signals in the Operating Room

J. W. R. McIntyre and L. M. Stanford

Anaesthetists must interact with apparatus to obtain information upon which much decision-making during clinical anaesthesia is based. The "anaesthetic machine" is such an apparatus, this term referring to a movable table to which are customarily attached gas and vapour delivery system, mechanical ventilator, scavenging and fluid suction systems, and monitors. In a previous study it was found that the relative positions of patient, anaesthetic machine, and anaesthetist adopted for a specific surgical procedure varied widely (McIntyre 1982). Often the anaesthetist faced the surgical site and the anaesthesia apparatus was behind him, an arrangement also described by Boquet et al. (1980). An implication is that under such circumstances the anaesthetist relied on an infallible ability to decide when it was necessary to turn and obtain information visually, or relied on auditory signals specified as alarms. These alarms are delivered from diverse equipment items made by different manufacturers and in the presence of environmental sounds. In this institution it was obvious that in the presence of certain hostile environmental sounds certain currently employed alarm signals could not be heard even if they were deliberately listened for. Thus it seemed important to study certain aspects of auditory alarm signals in the operating room with a view to improving our understanding and our arrangements. These aspects were their relative sites, origin and reception, their auditory characteristics and their interaction with certain other sounds in the operating room. This project involved a Boyle anaesthesia apparatus (Medishield) upon which were mounted an oxygen monitor (Ohio 200), blood pressure monitor (Dinamap 845, Critikon), transcutaneous PO_2 monitor (Kontron Medical 820) and airway pressure monitor (Ventilarm 5520, Ventronics). Included also were a minidrip (IVAC 530) and ECG monitor (Tektronix 412). The operating room measured 6.1×2.7 m and had ceramic-tiled walls, plastic-tiled floor and painted metal ceiling.

The project was conducted in three parts:

1. The question posed was: "If the alarm sounds were the same, or nearly so, would the spatial auditory discrimination of a healthy anaesthetist enable him to know which alarm was sounding?" To answer this, large-scale diagrams were made of the relative positions in the operating-room of the anaesthetist and the alarm signal sounds. The angle formed at the anaesthetist's head by sounds arriving from different directions was measured (Figs. 1 and 2) and comparison made with published data about spatial discrimination of auditory sounds. The signal sources could be well within a cone of auditory confusion. Thus under certain circumstances in the operating room, when the

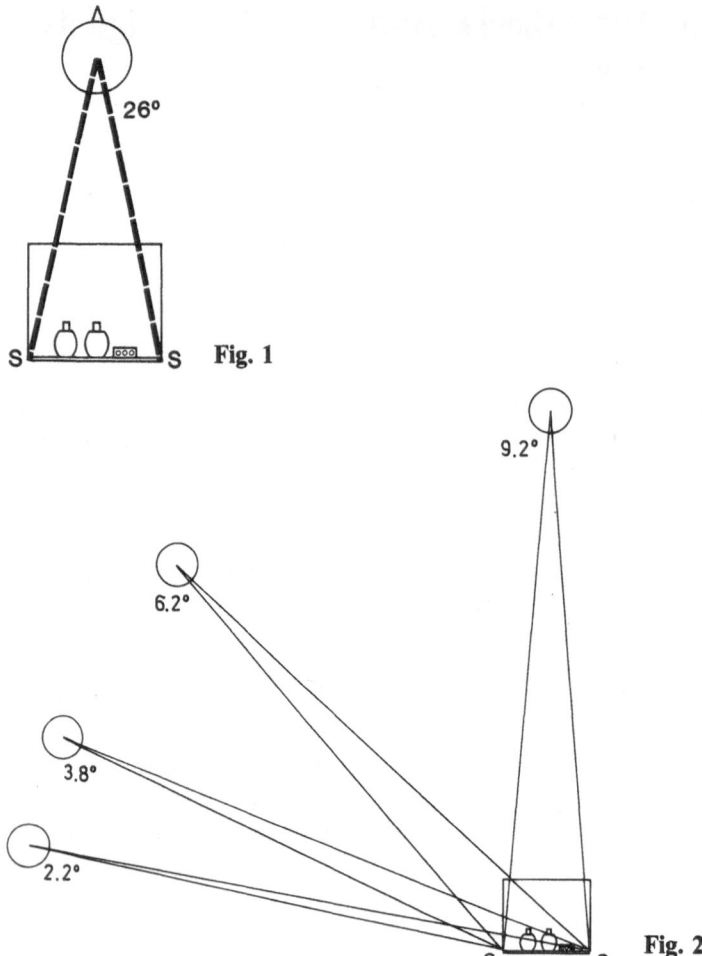

26°

9.2°

6.2°

3.8°

2.2°

S S   Fig. 1

S S   Fig. 2

anesthetist was moving about to perform some duty or maintain vigilance, he could not immediately and infallibly decide which alarm signal was occurring should the different instruments make the same or very similar signals. It must be noted that in the case of the ECG the alarm signal is the failure of the signal to continue its normal repetitive incidence.

2. The question posed was "Are there any similarities between the alarm signals customarily available in our operating-rooms?" A microphone (UHER 4000 – Report – L) was positioned at the site of the head of a seated anaesthetist at the head of the operating table, thus being approximately 1.5 m from the sources of alarm signals. Recordings were made on a tape-recorder (UHER 4000 – Report – L) at a time when the operating room area was not in general use. These recordings were subsequently analysed on a sound spectrograph (Sona-Graph 6061, Kay Electric) in the linguistics laboratory. The results of this analysis are shown in Table 1. All the alarm signals differ from one another to a greater or lesser degree. The relative amplitudes of the signals are microdrip > transcutaneous oxygen monitor > airway pressure monitor or blood pressure monitor > oxygen analyser > ECG.

Table 1. Analysis of operating room alarm signals

	Duration characteristics	Frequency (Hz)	Amplitude characteristics[a]
Airway pressure monitor (Ventilarm 5520, Ventronics)	Continuous tone	2800	Fairly steady amplitude 36 dB
Oxygen monitor (Ohio 200)	Intermittent tone, 660 ms in duration, repeated at regular intervals of approximately 2.5 s	Multiple components: 600, 2250, 2800, 3400, 4000, 4600, 5000	Random amplitude variations 32 dB (average)[a]
ECG (Tektronix 412)	Intermittent tone (analogue with stimulus from patient) 30 ms duration	2000	Low amplitude against some low-amplitude background noise 28 dB[a]
Transcutaneous PO_2 monitor (Kontron Medical 820)	Continuous tone	2900	Random amplitude variation (2–3 peaks per 100 ms) 36 dB (average)
Blood pressure monitor (Dinamap 845, Critikon)			
High tone	Continuous tone	Four components: 900, 1750, 2600, 3400	Steady amplitude 37 dB[a]
Alternating tone	Continuously repeated 240-ms cycle, during which frequency and amplitude of various tones change	Five components: 500, 800, 1200, 1750, 2600	Amplitude of 800-, 1750-, 2600-Hz tones decreases across cycle. Amplitude of 1200-Hz tone decreases from midpoint of next. Amplitude of 500-Hz tone has two high-to-low cycles within main cycle 32 dB average[a]
Low tone	Continuous tone	Four components: 1000, 1650, 2200, 2750	Steady amplitude 34 dB[a]
Minidrip monitor (IVAC 530)	Intermittent tone 400 ms tone 200 ms fade 800 ms silence	3200 with rise to 3400 in last $1/3$ of tone	Steady amplitude during tone 38 dB

[a] Amplitude of delivered signal can be changed by operator

3. The question posed was "Are any of the alarm signals masked by either of two instruments routinely used in our operating rooms – a gas-driven reamer-driver (Zimmer) or a high-volume dental suction apparatus (ADEC)?"
   A microphone was sited in the operating room equidistant between the alarm signal source and the surgical device (Fig. 3). This site for the microphone was

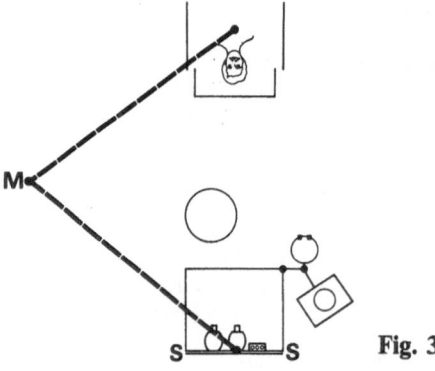

Fig. 3

Table 2. Analysis of hostile sounds

	Duration characteristics	Frequency (Hz)	Amplitude characteristics
Gas-driven reamer-driver (Zimmer)	Continuous random noise	1000–8000 (top of sonogram = 8000) Distinct tones at 1200, 2400, 3300, 3600, 4800, 6000 and 7150	Steady amplitude (est.) 37 dB
High-volume dental suction (ADEC)	Continuous random noise	80–8000 (top of sonogram = 8000). Wide spectral energy bands around 1000, 2500, 4200, 6050 and 7100	Steady amplitude (est.) 37 dB
Airway pressure monitor (Ventilarm 5520, Ventronics)	Continuous tone	2800	Fairly steady amplitude 36 dB

selected because anaesthetists quite often situate themselves at places in the operating room other than at the head of the operating table close to the anaesthesia machine. The results of the analysis of hostile sounds are shown in Table 2. The decibel values represent relative differences between the sounds and not real values. Comments on sonograms of hostile sounds in the presence of alarm signals are as follows:

a) Airway pressure monitor (Ventilarm) and reamer-driver: the 2800-Hz continuous tone of the Ventilarm can be seen on the sonogram between tones of the reamer-driver against a background of random noise. The greater amplitude and complex tone structure of the driver appear to mask it.

b) Oxygen monitor (Ohio 200) and reamer-driver: the oxygen monitor, because of its intermittent pattern and complex tone structure, is more clearly visible through the random noise of the driver.

c) Transcutaneous oxygen monitor and reamer-driver: the 2900-Hz continuous tone of the monitor can be seen on the sonogram between the tones of the drill and against a background of random noise, but the complex tone structure of the drill appears to mask it completely.

d) Airway pressure monitor (Ventilarm) and high-volume dental suction (ADEC): the 2800-Hz tone of the monitor can faintly be seen through the random noise but is almost completely masked by it.

c) Oxygen monitor (Ohio 200) and high-volume suction: the intermittent pattern and complex tone structure make the oxygen monitor more clearly visible through the suction's random noise than are either the airway pressure monitor or the transcutaneous oxygen monitor.

f) Transcutaneous oxygen monitor (Konitron) and high-volume suction: the 2800-Hz tone of the monitor can be faintly seen through the random noise of the suction but is almost completely masked by it.

It should be noted that these sonograms demonstrate similarities between physical characteristics of the sounds, and conclusions regarding comparative audibility are merely inferential. However, because of its complex tone structure the reamer-driver could completely mask the single-tone Ventilarm and transcutaneous oxygen monitor, even though their tones may be seen on the sonogram. The amplitude of the dental suction apparatus provides an almost complete auditory mask of the same two devices; they may be heard only faintly even if a deliberate attempt is made to detect them.

In discussing the results of the complete study consideration must first be given to the role of auditory alarm signals in the operating room during surgery. An alarm should be a warning of imminent danger for the patient, presented in a way that will infallibly attract the anaesthetist's attention. Anaesthetists' vigilance varies, influenced as it is by duration of work, attitude to the ongoing events, distractions, and diurnal rhythms. In addition, several vigilance tasks must be performed, and as these cannot be performed simultaneously time must be shared between them and the other activities anaesthetists engage in. One of the ways in which anaesthetists try to protect their patients and themselves against vigilance decrement is moving about the operating room and locating themselves a few metres distant from the anaesthesia apparatus. Indeed, such movement may also be necessary to discharge other duties. Thus continuous scanning for visual data is often inappropriate and auditory alarm signals of serious threats to a patient's welfare are necessary. Although every anaesthetist does his best to acquire relevant information before an alarm-sounding situation has been reached, once such a situation has been reached the alarm must always be detected immediately by the anaesthetist.

The study of individual alarm signals reported here (Table 1), their site of origin and the varying interaction between them suggests that a busy anaesthetist could not immediately distinguish between them, particularly if echo phenomena in the operating room had to be contended with. The interactive sound studies (Table 2) were conducted in the absence of any other operating room procedural sounds. It seems that in a working environment many unsuspecting anaesthetists would fail to hear some auditory alarm signals. Thus the conclusion is that the arrangements described are inappropriate for patient and anaesthetist.

A consideration of possible improvements must take into account the number of monitoring instruments from which auditory alarm signals are available or will be in the near future. There are likely to be nine of these. It is unlikely that their sites of origin can be conveniently arranged in compact anaesthesia apparatus so that the identity of individual signals can be determined on a directional basis at once by ear. Accordingly, each signal must be immediately identifiable by its auditory characteristics. An anaesthetist can easily learn to identify this relatively small number of characteristic signals, but as they occur infrequently the anaesthetist may easily forget their identity. Machine-talk, or in some instances machine-shout, will be one answer to this problem. Another may be a reduction in the number of possible auditory signals by designating one each for certain systems – inhaled vapour/gas information, cardiovascular information, and airway information. The alerted anaesthetist would then enter a decision tree including a search for further auditory and visual data about a particular system. None of these suggestions address the vital question of interaction between auditory alarm signals and environmental noise in the operating room. Increasing the amplitude of the delivered signal is unduly disturbing to the operating room personnel in the absence of the hostile sounds. We are at present addressing this problem by designing alerting signals on the basis of the hostile sounds with which they must contend. However, until such time as appropriate signals are in use, the careful anaesthetist must remember to concentrate deliberately on visual data acquisition when hostile noise occurs in the operating room or any other working environment for patient care.

Acknowledgement. Allen J. Opperthauser, Technician, Department of Linguistics, University of Alberta.

References

Boquet G, Bushman JA, Davenport HT (1980) The anaesthetic machine – a study of function and design. Br J Anaesth 52:61

McIntyre JWR (1982) Man-machine interface: the position of the anaesthetic machine in the operating room. Can Anaesth Soc J 29:74

Rapid Setting of Intravenous Infusion Rates

P. A. Foster

One of the most frequently performed tasks in intensive care and postoperative wards is supervision of intravenous infusions. The ideal of an electronically regulated infusion pump for each patient is seldom achieved, or indeed justified for routine care. However, a work study of a nurse delegated with supervision of intravenous infusions will show that:

1. The setting of an infusion drip rate, if this is not done by guesswork, takes at least 2 min. The slower the rate the longer the time taken.
2. One reason for the length of time needed is that the rate is counted against the second hand of a watch or digital stop-watch, looking from one to the other. The clock and the drip chamber are never juxtaposed.
3. In the conventional intravenous infusion set with roller clamp flow regulator, the drip rate tends to slow down with time, partly because the hydrostatic pressure falls, and partly because the interposed flow resistance (the roller clamp) rises as the plastic deforms under pressure. One thus needs to check and reset every hour.
4. The nursing duties involved in immediate postoperative care include the ¼- or ½-hourly checking of pulse, blood pressure and respiration, and possibly the keeping of a related pain control chart. These routines may clash with the intravenous supervision routine.

In terms of nurse hours per day, the supervision of say ten postoperative patients after routine uncomplicated surgery may be a full-time task for one nurse. In an attempt to make this routine and daily task more efficient, attention was paid to the one duty seen to take most time and to be done most unsatisfactorily – setting the drip rate. A small, cigarette-packet sized unit was designed in which a visual reference rate could be set and held adjacent to the drip chamber with one hand, whilst the other hand manipulated the roller clamp so that the fall of the drip coincided with a high-intensity short-period flash from a light-emitting diode.

This unit was tried with subjects totally unfamiliar with intravenous administration procedures. After a few minutes familiarization with the equipment, these persons could accurately set a drip rate within as little as 10 s with a mean time of about 25 s – depending somewhat on the rate of infusion. It is easy to incorporate into the design a 15- or 30-s timer which the nurse can use in counting pulse and respiration rates, thus allowing several tasks to be performed at each visit to the patient. The cost of the unit is approximately one-tenth that of an infusion pump. It can, of course, be used to regulate a number of infusions simultaneously.

The reference rate can be set between 20 and 100/min, which meets most clinical situations, corresponding to an infusion rate of 1 or 2 to 5 or 10 ml/min with adult sets. Rates outside these limits can easily be set by using a 2:1 ratio between reference and drop or a set giving 60 drops/ml.

The unit is available from B. Braun, Melsungen, or from: Art Medical Equipment, P. O. Box 23591, Joubert Park, 2044 Republic of South Africa.

Presentation of an Ergonomically Designed Work Place for Anaesthesia

R. Droh

For 20 years I have been perplexed and irritated by the inconvenient and cumbersome design of our various instruments; by their poor arrangement; by the fact that cables are left trailing and extra supplies standing around, with many obstacles on the floor likely to cause an accident; by the instability of infusion stands and other equipment; and by poorly fitting collectors and couplings and poorly adaptable connectors. Many of our instruments are still impractical, too delicate, too heavy, too expensive, time-consuming and sometimes enervating. In the past 10 years all these factors have driven me to design and construct my own new work places.

Figure 1 shows one such mobile work place. The most important thing about them is that we build chassis with three, four, five or six wheels and construct on them a seat for the anaesthetist together with a place for instruments and/or respirators, a movable wing for the whole anaesthesia machine, another movable wing for the monitoring equipment and a movable desk. All these different pieces are exchangeable parts of the whole carrier. It is thus always possible to put together a practical and mobile work place which meets the particular demands of the situation. Cables and gas supply pipes are safely enclosed, but are still easily accessible for repair. With this combined, movable work place it is not necessary to search for a stool, infusion stands, instruments, anaesthesia machine, monitors etc. Incidentally, all the instruments are arranged in an efficient order in a minimal space. The contact between the patient and the instruments can be made with a few manipulations. The work places in the different operating theatres are thus

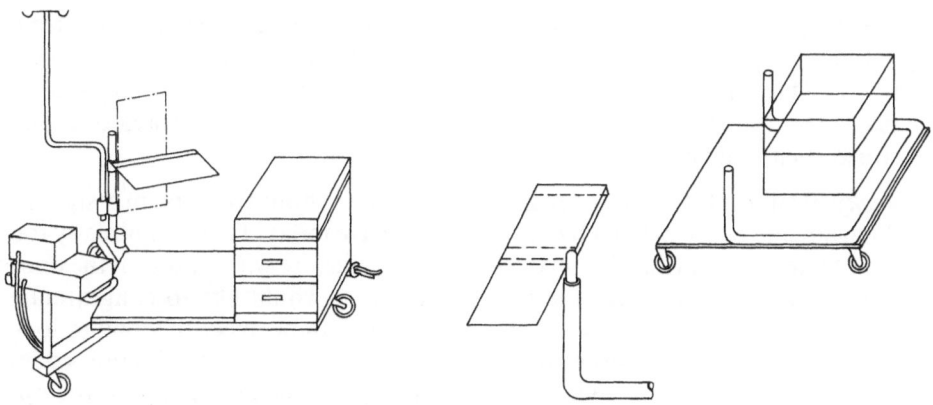

Fig. 1. The newly designed mobile anaesthesia work place

very well standardized but can also be efficiently adapted to the requirements of each particular case.

The sitting, writing and work surfaces are all practically and comfortably located in relation to one another. It is impossible to knock over a stool, infusion stands or other instruments. The position of the anaesthetist in relation to the various machines can be so arranged that all parameters are easy to observe, measure, and adjust. The arrangement of the various surfaces for the various instruments is adaptable as far as the instruments of the different companies allow. Of course, we are still dependent to some extent on the shape and size of the instruments.

References

Droh R: Fahrbarer Geräteträger. Deutsches Patentamt G 76 40 583
Droh R: No 7731053 Patent Office, Paris
Droh R: No 474 25/77 Patent Office, London

Discussion V

Erdmann: We want to have a better control of our patients. Our material should be better at hand. We want to have ergonomic workplaces.

Foster: I think it is very important to have auditory signals, because you must listen to your machines. That is one of the reasons why we don't have music in our operating-theatres. For the patient yes, but not for the anaesthetist. There are, of course, different ways of coding signals, for example with pure sinus waves with different frequencies between 300 and 3000 Hz. This is the telephone spectrum. You can have different grades of alarm signal, pure sinus wave as your first alarm and a modulated sinus wave as your second alarm signal. You can have non-emergency signals at a low frequency indicating that the normal function is being performed. There are many ways that one can do this. I think there should be an international standard.

One thing that I did not mention before: nowadays we have standard fluid crystal monitors, which are very compact and which can be driven by a microprocessor. This monitor can be mounted next to the patient, so you can have the feedback from your patient next to the patient's head. The Havard group described such a device.

Siedlecky: Prof. McIntyre stated in his lecture that we should develop new signals. But I am already overwhelmed by the signals we have today. There are already too many of them. Sometimes I dream of an operating-theatre with a note on the door outside saying, "Silence, anaesthesia in progress." I would like to congratulate Dr. Droh on the organization in his clinic. He has a simple piece of wood to lock the door of the induction room. In my hospital we have a new induction room now. Everybody passes through the induction room when he is going to the operating-theatre. And this happens exactly at the time of induction.

In Augsburg I have a telephone on every anaesthesia machine. This is a nightmare. When I analysed the messages coming through this telephone I saw that one-third were totally irrelevant for our work in the operating-theatre. Private telephone calls to the surgeon, irrelevant talks between sister and surgeon etc. I always say, "Excuse me, but you are disturbing my work." Every one of my colleagues spends a lot of time answering such telephone calls. So I can only warn you all not to have a telephone installed on your anaesthesia machine.

Then we have a new ECG signal and everyone was happy to have it. But what does this signal show us? A normal activity. So I always turn this signal off nowadays. I want to have an arrhythmia alarm and nothing more. Why should we watch a normal cardiac activity? This is only acoustic pollution.

We have signals which are very loud, but you cannot change the loudness. Some of our signals make the surgeons very nervous and they ask us to stop the noise, but we cannot do so. Besides we have music in the operating-theatre. But it is impossible to satisfy everyone's taste.

Then you have your emergency bleeper. Sometimes we have two of them: one Eurosignal and one clinic duty signal.

Another problem is the mobility of the anaesthesia machines. The machines are too heavy. Also, those hanging from the ceiling are too heavy. Do they have to be so immobile?

Droh: I would like to underline what Dr. Siedlecky said. We have some alarm signals which give alarms just when you don't need them. In the future we must have the possibility of reducing the number of these different alarm signals.

One word about the heavy machines. I have always said that these are "dinos". They ought to be lighter. Perhaps one could try using special plastics instead of steel.

Siedlecky said that the he was impressed by a simple piece of wood blocking the door to the induction room. I think we should always try to find a simple solution. The industry should not build more and more highly sophisticated anaesthesia machines with electronics and automatic gas mixture. It is nearly impossible to change such a machine yourself. But we must have the opportunity to realize new ideas. I would never have dared to use closed-circuit anaesthesia with these new machines.

One word about the German Society for Anaesthesia: they are always establishing new technical rules blocking our creativity. They should be careful not to stand in the way of progress.

Meyer: I think we should look for the reasons for those highly sophisticated anaesthesia machines in ourselves. All anaesthetists are more or less pseudomechanics. But I think it is the task of the industry and not of an anaesthetist to improve our equipment.

Also, the preferences of anaesthetists are quite different. Many of us want to have a computer steering and controlling anaesthesia, so that we only have to look at the control signals. We should tell the industry to build machines that are less heavy but more effective. We want to perform reasonable anaesthesia and we want to do something for the patient and not to fight with our machines. We no longer want to fight with the surgeon and to hide behind big machines. Some time ago we needed big machines to hide behind. As long as we stick to big machines as a sign of our status, we will have a Mercedes and not a Volkswagen which would be as efficient as the Mercedes. It is a pity that the organization responsible for the design of anaesthesia machines has no common idea of such a machine. Everyone wants to realize his own idea. It is useless to discuss this subject further as long as somebody in Ulm decides what an anaesthesia machine should look like.

Droh: There is something in that but I cannot agree with all your statements. I don't think it is right that we needed big anaesthesia machines to hide behind. In those days nobody was able to build small machines. In the mean time micro electronics has changed many things. But the anaesthesia machines should be as easily handled as a new car. They should be light and easy to reconstruct and use new technology. Every anaesthetist should in a small way be a mechanic too. Anaesthesia is the combination of technology and medicine.

We still have no good technical rules and standards. On the other hand, some time ago we had machines which were realy dangerous. But in the mean time there are so many different rules that the industry is hardly able to change the technical conception of its machines. Our symposium should help to bring together those making the rules, those having the ideas and those producing anaesthesia machines.

Foster: I think there are three problems. The first is the problem of ergonomics. In my experience many doctors don't know how their machines work. I have been involved in the design of many operating-theatres, and the first thing that I always noticed was that the doctor was not able to read the plan. For example, the anaesthetist does not know from which point to which point he walks every day, where the patient comes into the operating-theatre and where he goes out of it. Dr. Siedlecky gave us the example of his induction room. In my opinion you have to talk to the architect about the design of the operating-theatre.

The second problem is the problem of sensory pollution. For example, we don't talk about light pollution, about differences in light concentration in the operating-theatre, about the differences in colour and temperature. It may be difficult to see what colour the patient is. Where we do have the right light-contrast, what amount of noise do we have? One needs to have standards for all these things.

McIntyre: There are some other aspects of what we are talking about to which I would like to draw your attention. There are some people who feel that the Volkswagen is a very good car. I possessed a Volkswagen and it was the most exciting and dangerous vehicle I have ever owned. I don't want to anaesthetize my patient with a Volkswagen but rather with a Mercedes.

The second point which I would like to make is that I agree whole-heartedly with the people who say we have too many signals. We have to restrict this to about three or four important signals.

One most important point is the cost of our monitoring-devices. We are nowadays forced to decide what monitoring we need to get an appropiate outcome for the patient.

A Newly Designed Unit for the Recording of ECG, Temperature, Pulse Plethysmography, and Other Body-Surface Parameters

R. Droh

This unit complements the ergonomically designed anaesthesia work place and simply needs to be placed under the patient in the appropriate position. It is a wedge-shaped cushion filled with foam rubber. The electrodes for the recording of ECG, temperature etc. are fitted into the material of the cushion. The whole is covered with an antistatic material which is easy to clean and to disinfect. The surfaces of the electrodes are welded into contact with the cover. The electrodes are pressed against the body of the patient by his own weight, ensuring optimal connection.

There is a cavity in the filling material of the cushion in which a plethysmographic pulse detector can be installed; this is then connected to the single cable. There is only one cable between the cushion and the monitors. Inside the cushion are connections between the ECG electrodes, thermoelectrodes, a large neutral electrode for added protection in the case of high-frequency surgery, a pulse detector and, if desired, other surface-parameter-measuring devices. Since the various elements are built into a cushion of the size and shape that the anaesthetist needs in any case to place the patient in the optimal position, work is spared. Further work is spared by the fact that there is only one cable with one plug; there is no need to connect several different cables to the monitors.

For the plethysmographic finger pulse there is a special finger attachment made out of foam in which the pulse detector is fitted. This, again, has only one cable,

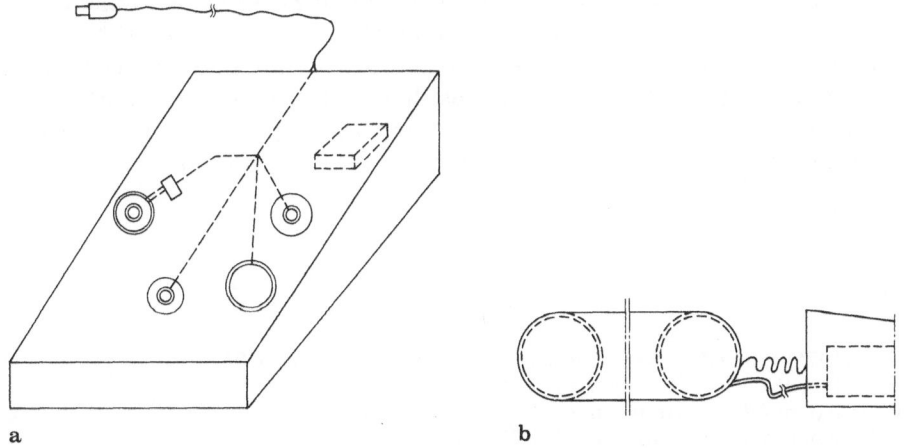

a

b

Fig. 1. a Unit for the recording of ECG, temperature, pulse plethysmography and other body surface parameters. **b** Cable box attached to the back of the cushion

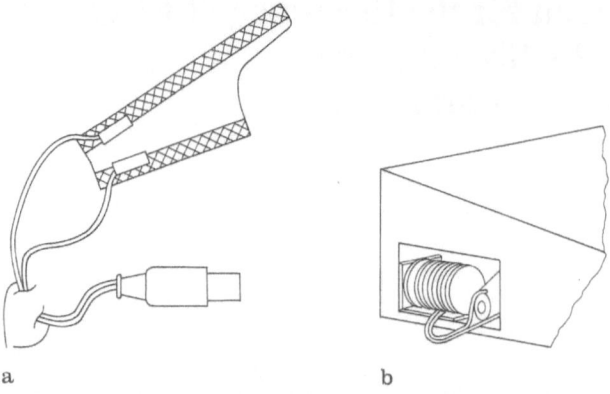

a   b

Fig. 2. a Finger attachment for pulse plethysmography. **b** Cable box for finger attachment

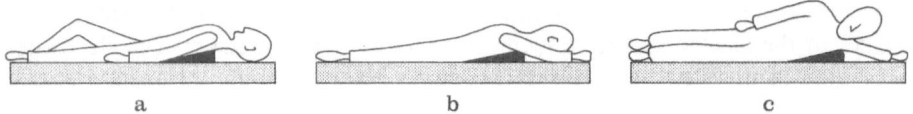

a   b   c

Fig. 3. Application of the ECG cushion electrode (**a**) in the supine position as with a normal shoulder support, (**b**) in the ventral position, and (**c**) in the side position

which is connected to the cable from the cushion and from there carries signals to the monitor. It is only necessary to draw the cable out of the cushion and place the finger attachment on the patient, and it is already connected to the monitor.

Foam rubber was chosen because it is soft and elastic but retains its shape. It scarcely influences the circulation of the blood in the finger. Rings of rigid material constrict the circulation to some degree. The material also adapts well to variations in finger length or breadth. The finger attachment makes it possible to measure very slight pulse and low blood pressure without being affected by auscultatory difficulties. It makes it possible to measure the peripheral arterial blood pressure curve, which gives us information about the functioning of the windkessel. With this attachment alone it is possible to visualize pulse frequency and rhythm, and thereby to determine otherwise hardly measurable blood pressure.

This unit is an integrated part of the anaesthesia work place. Figures 1–3 show the unit together with the finger attachment and their application.

References

Droh R (1975) Eine EKG-Kissen-Elektrode. Anästh Praxis 11:19–21
Droh R: Patent Office, Munchen, G 73251419
Droh R: Patent Office, Munchen, GM 7615539
Droh R: Patent Office, Munchen, GM 7611383
Droh R: Patent Office, London No. 1579376
Droh R: Patent Office, Paris 7710983

Discussion VI

Siedlecky: Dr. Droh do you always get a good ECG signal, especially when the patient is lying on his side?

Droh: Yes. Sometimes we have some problems with high frequencies coming from the electric cauter, but the new machines have better filters.

The Organization of the Secretariat

S. Gonserowski-Spintge

In a 300-bed clinic for orthopaedic surgery we undertook a study in the secretariat of the chief anaesthetist in order to improve its organization. The main achievments were meant to be:

To raise the output while making the course of work more comprehensive and more rational
To save time
To save money.

Certain points had to be recognized:

With effect from January 1st 1983 there are new rules concerning the payment for treatment of private patients in the Federal Republic of Germany.
A physician must nowadays be a manager as well.
The chairman of a department should as far as possible be free from administrative tasks in order to have more time and energy to take care of his patients, to do research, to inform himself.
There should be a middle way between too much organization and too little rationalization.

Method

To achieve these aims there must be a balance between:

The equipment (typewriter, personal computer, telex, telephone etc.)
The surroundings (design of the secretariat, climate, light etc.)
The tasks
The qualifications of the secretary (training, personal interests etc.).

At the beginning we established a 1-week time-study as an analysis of the present situation. All work was recorded in detail and the time necessary to fulfil every task was measured.

Results

From these data we were able to find out main tasks and weak points (Table 1). We designed an ideal work programme and planned how it could be achieved.

Table 1. One-week time-study

Tasks	Weekly time		Weekly money (DM)
	(min)	(%)	
Preparation of the bureau	81	3.5	52.50
Telephone calls			
Secretary himself (number: 72)	57		
From outside (number: 89)	61	5.1	76.00
Personal contacts (whole number: 115)	134	5.7	85.50
with the chief only (number: 40)	123	5.3	79.50
Handling of letters	164	7.0	105.00
(Receive, open, discuss with chief, send out etc.)			
Answering letters	171	7.3	109.50
Dealing with invoices	48	2.1	31.50
Liquidation (patients with private health insurance or none) (invoices, letters requesting payment etc.)	655	28.1	421.50
Chief correspondence (scientific, private)	258	11.1	166.50
Preparatium of information and papers for the chief	46	2.0	30.00
Preparation of papers for the tax expert	36	1.5	22.50
Symposia	45	1.9	28.50
Journals (establish bibliography)	50	2.1	31.50
Organization of the secretariat	37	1.6	24.00
Training	33	1.4	21.00
Taking care of assistants and trainees	13	0.6	9.00
Documentation	38	1.7	21.00
Transfering messages and other special duties	282	12.1	181.50
Total	2332	100.0	1500.00

Having discussed the aim and the plan to realize it with the chief and the secretary, we established a catalogue of means and actions together with a timetable for realization (Table 2).

Table 2. Catalogue of planned improvements

Improvements	Realized by
1. Introduction of a personal computer	15. 5. 1983
2. Special training for the secretary	2.–4. 6. 1983
3. Date stamp	1. 5. 1983
4. Colour code for notes: red = telephone green = information from the secretary	1. 5. 1983
5. Every Friday control of dates and time table and date-time papers (Papiere mit Wiedervorlagefrist)	1. 5. 1983
6. Noting far-ahead dates in the diary	1. 5. 1983
7. Address file for Christmas, birthdays, special occasions	1. 6. 1983
8. Document file for the daily photocopies	1. 5. 1983
9. Delegation of transmission of messages	15. 5. 1983
10. Postal franking machine	15. 5. 1983
11. Check-list for symposia and travelling	30. 6. 1983

(will be continued)

Table 2 (contd.)

Improvements	Realized by
12. Documentation for the library	1. 5. 1983
13. Documentation of papers given away	15. 6. 1983
14. Documentation of all files (code-number, place)	30. 7. 1983
15. Monthly financial check-up	1. 5. 1983
16. Documentation of invoices on microfiche	1. 7. 1983
17. Discussion of organizational problems	
Chief – secretary (once a year)	18. 5. 1983
Chief – staff (twice a year)	5. 1983 + 11. 1983
18. Office hours for the secretary	1. 5. 1983
19. Calculator	15. 5. 1983
20. Telephone with indication for internal/external calls	1. 1. 1984
21. Delegation of "personal contacts" to the 2nd secretariat	1. 5. 1983
22. More time for the trainees and assistants (feedback later)	1. 5. 1983

Consequences and Discussion

In this secretariat we were able save 164 h/year through the realization of the above-mentioned catalogue of improvements. Looking at this result one must take into consideration that the secretary already used an "intelligent typewriter" with a mini floppy disc. However, to save 164 h means:

To save 4 weeks a year
To reduce the weekly working hours from 40 h to 36.5 h and increase the working capacity of 1 year by 8%
To save about 6000 German marks (= $2400) in salary each year.

On the whole, using an intelligent type writer or personal computer it is even possible to save 1700 h (= 42 weeks = 36 000 German marks = $ 14 000) per year.

Investment in a personal computer (about 20 000 German marks) would be recouped within 1 year.

Discussion VII

Erdmann: Thank you very much for this most interesting paper. I think I will have to reorganize my whole office.

Droh: I would like to underlinde that our secretary should have a double training. First she should be a nurse or a medical assistant, and secondly she should be a well-trained secretary. I am lucky to have such a secretary. She can work as an anaesthesia nurse in the operating-theatre, she can work in the laboratory and she can work as a secretary. I think it is essential that she knows all the work of her chief.

In my office I have two light signals. When both of them are on I don't want to be disturbed at all. If one light is on I only expect to be disturbed by very important things. If no light is on anyone can disturb me.

In this connection we should also know how much work we can give to our secretary, how we can help her to organize it better etc. We should always consider that it is a case of *Do et das* (I give and you give).

Erdmann: I completely agree. Prof. Oyama, how is the secretariat organized in Japan?

Oyama: The Japanese system is a litte bit different from the German system. The German system seems to be better organized. The Japanese secretary has to fulfil all those tasks which Mrs. Spintge mentioned before and more besides.

I also agree with Dr. Droh that our secretary needs some recreation. We should give her between 30 min and perhaps 1 h free time each day. That is my own personal experience.

Droh: How long does the secretary work per day in Japan?

Oyama: In my department the secretary arrives at 8.30 in the morning and she goes home at 6.00 o'clock in the evening, sometimes even at 7.00 o'clock. For instance, when I have to prepare for a symposium like this one she must work at the typewriter until my paper is finished.

Droh: How many days per week does she work?

Oyama: Six days: Saturday is half a working day.

Dr. Siedlecky: I would like to comment upon the tasks of a secretary in a teaching hospital. A typical situation is that new doctors or new students come to the department at about 8.00 o'clock in the morning when the doctors are already in the operating-theatre. I think that the first impression you get when you come to a new department is very important. When I came to this hospital in Lüdenscheid my first impression was very positive because people were very efficient: they welcomed me at once and brought me to the right place. In our clinic we have the problem that on Mondays 20 students come for their practical course. In Boston every guest got a plan of the hospital, there was a special secretariat for the welcome of guests, and the secretary even helped you to search for a flat in the town.

Erdmann: In Rotterdam this is different. The chief secretary in a big department cannot take care of every guest. Different people have to do different jobs there.

Droh: Right, we should beware of overloading our chief secretary. However, one can mention that in Japan a secretary has to work 55 h/week. But as Prof. Erdmann said, we have to have different secretaries for different tasks. Nevertheless, one must be sure that the chief secretary knows what is going on. Otherwise, one is not informed oneself in the end.

Mrs. Gonserowski-Spintge: In our study we saw that more work can be done in less time when there is the right organization. Today there is no longer a linear relationship between working time and work-output. To achieve this you must look at your secretariat, look at the organization, at the amount and the character of work, at the working-tools which are used. What I told you is not the general scheme for every department in every clinic. You must look at your individual tasks and think what could be improved. Everywhere there are possibilities of making work more rational and more human.

The Induction and Recovery Period
of Inhalation Anaesthesia:
A Considerable Improvement with Isoflurane

B. Dworacek and E. Lachitjaran

Isoflurane was administered to 60 patients during different surgical procedures. Aspects of inhalation induction in 35 patients and the recovery period in all 60 patients were studied. These aspects seem to us to be of importance in surgical emergencies, in high-risk patients, paediatrics, and also in out-patient anaesthesia.

With halothane the induction is rapid and easy. The relaxation of the lower jaw muscles is adequate if endotracheal intubation is to be performed. The low minimal alveolar concentration (MAC) value of halothane (0.75 vol.% in oxygen, 0.29 vol.% with nitrous oxide 70%) and the use of halothane vaporizers with a maximum of 5 vol.% are essential for a rapid induction. The recovery time (the time between the end of inhalation of the agent and arousal) with halothane anaesthesia is conveniently short and waking is seldom followed by nausea or vomiting. The incidence of the central cholinergic blockade (central anticholinergic syndrome, CAS) does not appear to be high in anaesthesia with halothane (Lommers et al. 1980; Dworacek et al. 1982).

Nevertheless, the halothane action on the heart and dysrhythmias bearing a direct relationship to hypercapnia (which easily occurs during the spontaneous breathing of the inhalation induction) make the use of halothane in high-risk patients dubious. The rather high biotransformation of halothane (15%–20%) is another disadvantage. Enflurane seems to have some advantages over halothane in general anaesthesia: low biotransformation rate (2.5%), an excellent muscle relaxation and a shorter recovery time.

However, inhalation induction with enflurane appears to be more difficult than with halothane. A high MAC value (1.68 vol.% in oxygen or 1.15 vol.% in nitrous oxide 70%) with the maximal vaporizer concentration of 5 vol.% is not beneficial for the inhalation induction. With enflurane, stage 2 is often accompanied by vigorous excitement. Signs of the CAS, i.e. delirious behaviour, shivering, convulsive movements or confusion accompany the arousal from an enflurane anaesthetic more often than that from halothane.

Methods

Inhalation induction with isoflurane was performed in 35 patients. Of these, 28 patients were endotracheally intubated as soon as stage 3, plane II was reached; in seven patients anaesthesia was further maintained with a face mask (under spontaneous breathing). All patients were premedicated 1 h before commencing anaesthesia with atropine sulphate 0.007 mg/kg; males ranging from 20 years to 60 years were given an additional 75 mg meperidine, females 50 mg meperidine i.m. Patients were preoxygenated for 1 min; a mixture of 80 vol.% nitrous oxide and 20 vol.% oxygen followed. When the first central effects of nitrous oxide (euphoria or loss of consciousness) were noticed (after 2 or 3 min), the isoflurane vaporizer (Fortec, Cyprane) was turned on at 1 vol.% isoflurane. After ± three breaths, the concentration was increased to 3 vol.% and subsequently to 5 vol.%. This concentration was maintained until the endotracheal intubation or until the stage 3, plane II was reached (patients on mask only). One minute before the endotracheal intubation, the larynx and trachea were sprayed with 3 ml 4% lidocaine (plastic cannula, Equip).

The time between the start of the isoflurane inhalation and the endotracheal intubation (or stage 3, plane II) was recorded as the "inhalation induction time." The following parameters were screened during the inhalation induction: the behaviour of the patient (especially signs of excitement in stage 2), respiration, heart rate, ECG, blood pressure (Riva Rocci, oscillometry or arterial pressure line).

Twelve patients breathed spontaneously during the maintenance of anaesthesia, which consisted of a mixture of nitrous oxide and oxygen 2:1 and 2.5–3.5 vol.% isoflurane. A non-rebreathing system was used (AMBU E-valve). Forty-eight patients were control-ventilated with a mixture of nitrous oxide and oxygen 2:1 and isoflurane 0.5–1.5 vol.% during the maintenance of anaesthesia. Pancuronium and fentanyl were additionally given; a circle system with CO_2 absorption was used.

The recovery time was recorded from the ceasing of isoflurane and nitrous oxide (vaporizer closed, patients breathing air spontaneously or ventilated with air after disconnection of the circle system) up to the moment that the patient was completely awake. Heart rate, blood pressure, breathing volume, and general behaviour were monitored.

Results

No patient complained of an unpleasant smell when isoflurane was preceded by nitrous oxide. The mean induction time was 4 min 15 s ($n = 35$, SD 1 min 18 s, range 2 min 30 s to 10 min). In one patient only some muscle movements were seen, reminiscent of a mild excitement. Others reached stage 3, plane II calmly, breathing rhythmically and deeply. There was no significant blood pressure fall or change in the heart rate. The relaxation of the lower jaw muscles was very good. The endotracheal intubation was easy and did not cause coughing when preceeded by the lidocaine tracheolaryngeal spray. The end-expiratory CO_2 did not exceed 5.6 vol.%.

The mean recovery time was 8 min 8 s ($n = 60$, SD 2 min 30 s, range 4–13 min). Recovery and waking were surprisingly calm in all patients. Patients opened their eyes and suddenly regained their contact with the surroundings. No confusion was seen, no excitement (emergency) or any other symptom of the central cholinergic blockade (CAS). None of our patients had nausea or vomiting. During recovery from isoflurane the blood pressure was stable; there were no dysrhythmias. The ventilation was sufficient in patients who had been breathing spontaneously. No problems were encountered in those patients who had been ventilated, and in whom relaxation was reversed by neostigmine. The earliest need for postoperative analgesia was recorded 30 min after arousal.

Discussion

Isoflurane is an isomer congener of enflurane. However, it has some other properties not seen in enflurane which are considered advantageous: the biotransformation is the lowest of all known halogenated volatile agents (0.2%). It does not cause "spikes" on EEG during anaesthesia and cardiac performance remains remarkably stable (Eger 1981). Our overall impression of these 60 patients confirms these findings and we feel that isoflurane is a welcome volatile agent, especially for inhalation induction.

There will remain a great need in the future for such an agent because of the intrinsic safety of inhalation anaesthesia. The virtual absence of excitement in stage 2 makes isoflurane very safe for the induction of anaesthesia in high-risk patients. The surprisingly calm recovery and arousal are of essential benefit, especially in out-patient care. The smooth induction of anaesthesia with isoflurane and the calm recovery suggest that isoflurane could be an agent of choice in disaster medicine.

References

Dworacek B, Rating W, Rupreht J, Lommers C (1982) Die Therapie des zentralen anticholinergischen Syndroms. In: Stoeckel H (Hrsg) Das zentral-anticholinergische Syndrom: Physostigmin in der Anästhesiologie und Intensivmedizin. Thieme, Stuttgart, S. 42–51
Eger EI II (1981) Isoflurane (Forane). A compendium and reference. Library of Congress Catalog Card Number 81-65686 Airco, Inc
Lommers C, Rupreht J, Dworacek B (1980) Physostigmine and its importance to the anaesthesiologist. In: Rügheimer E, Wawersik J, Zindler M (eds) Abstracts of the 7th World Congress of Anaesthesiologists, Hamburg, Excerpta Medica, Amsterdam, p 209

Isoflurane

G. Rolly, L. Versichelen, R. Serreyn, and M. Dhont

Isoflurane, the isomer of enflurane, is a novel inhalation anaesthetic. The most attractive feature is certainly the low degree of biotransformation, which is lower than that of any of the currently used potent inhalation anaesthetics. The minimum alveolar concentration (MAC) value is around 1.3% in the young age group; the blood gas partition coefficient of isoflurane is 1.4 (compared to 1.9 for enflurane and 2.3 for halothane), which is the main reason for the rapid induction of anaesthesia.

As with other potent inhalation anaesthetics, respiratory depression occurs in spontaneously breathing patients; it is somewhat less if N_2O is added. An attractive feature of isoflurane is also the lower tendency to develop arrhythmias if adrenaline is used concomitantly. Compared to the other potent inhalation anaesthetics, isoflurane produces a more important decrease in peripheral vascular resistance and in blood pressure, but in contrast cardiac output is well maintained due to the mild increase in heart rate.

At the University of Ghent we used isoflurane as an early clinical trial in two groups of patients: a) a group of female patients anaesthetized for hysterectomy, in whom it was used as routine clinical anaesthesia; and b) a group of patients in whom it was used for a special closed-circuit anaesthesia technique, by injection of liquid isoflurane into the circuit.

Isoflurane Anaesthesia for Hysterectomy

In Table 1 the anthropometric data for the first group of ten hysterectomy patients are given. The patients were premedicated with atropine 0.5 mg and Thalamonal (droperidol plus fentanyl) 2 ml/70 kg. An intravenous infusion with NaCl 9‰ was performed. The patients were preoxygenated with oxygen (6 l/min) given by a plastic face mask. An etomidate drip (50 μg/kg per minute) was started to ensure basal sedation of the patient while the arterial sampling catheter was

Table 1. Isoflurane $1.3\%/O_2/N_2O$ Anaesthesia in ten female patients

Age	44.7	±2.7	Years
Weight	62.8	±1.7	kg
Height	161.1	±1.6	cm
Body surface	1.655±0.017		m^2

introduced percutaneously (Angiocath). Anaesthesia was induced with etomidate 0.3 mg/kg and muscle relaxation obtained with pancuronium 0.1 mg/kg. After 2–3 min the patients were intubated and ventilated by means of an Engström respirator used with an open system. Isoflurane 1.3% was given at a constant inspired concentration in a 50% O_2/50% N_2O mixture. No analgesic supplement was added.

Although monitoring was continuous, measurements were made at several standardized examination points: I, awake; II, after anaesthesia induction; III, 5 min, IV, 25 min, V, 50 min and VI, 75 min after surgical incision; VII, just after awakening; VIII, 60 min after awakening. At these exact moments, systolic and diastolic blood pressures, heart rate, stroke volume and cardiac output (pulse contour method) were recorded. Blood was sampled for blood gas measurements (PaO_2, pHa, $PaCO_2$, base excess, standard bicarbonate) and for determination of glycaemia and of plasma levels of cortisol. The open anaesthesia system was temporarily converted to a true closed system in order to measure oxygen consumption by a special technique using the EHN spirometer. Reversal of muscle relaxation was done at the end of anaesthesia by injection of neostigmine 1.25 mg combined with atropine 0.5 mg. The duration of anaesthesia was 160.5 ± 13.5 min (mean ± SEM) and that of surgery 132.5 ± 12.6 min.

Results

The cardiovascular parameters are shown in Table 2. Although systolic blood pressure was lower than before anaesthesia, only at two points was the decrease statistically significant. On awakening, systolic blood pressure was higher than before anaesthesia induction. Diastolic blood pressure did not change significantly from one examination to another. Heart rate increased significantly throughout anaesthesia and decreased after awakening. Stroke volume decreased somewhat in the course of anaesthesia and was higher on awakening than before anaesthesia. However, cardiac output changes were never statistically significant.

A statistically significant increase in PaO_2 occurred when ventilation of the patients with 50% oxygen was started; before the anaesthetic the patients were breathing only oxygen by mask (232.6 ± 12.7 vs 156.7 ± 33.8 mm Hg); PaO_2 remained increased throughout the examination period. pHa changes were only statistically significant at some points and were subsequent to a slight decrease in arterial $PaCO_2$ (I, 44.2 ± 1.5 mm Hg; V, 35.7 ± 1.6 mm Hg) due to a moderate hyperventilation and a moderate decrease in standard bicarbonate (I, 24.34 ± 0.59; V, 22.04 ± 0.33 mEq/l) and in base excess, which were at some points statistically significant.

Some metabolic changes are shown in Table 3. Throughout the period of anaesthesia, oxygen consumption decreased as time went on, probably reflecting a somewhat deeper anaesthetic plane at the constant inspired concentration. Glycaemia increased statistically, although the patients received only an infusion of normal saline. The cortisol levels show a constant decrease throughout anaesthesia and in the postoperative period; several times this decrease was very significant.

Table 2. Cardiovascular parameters ($n=10$)

	Examination points							
	I Awake	II Anaesth.	III Surg. +5 min	IV Surg. +25 min	V Surg. +50 min	VI Surg. +75 min	VII Awake	VIII Awake +60 min
Syst. B.P. (mmHg)	141.8 ±3.4	132.2 ±5.3	139.4 ±10.1	128.7 ±6.6[a]	124.5 ±6.4	116.8 ±6.1[b]	149.7 ±5.8	139.7±5.0
Diast. B.P. (mmHg)	82.3 ±2.8	76.4 ±5.5	82.3 ± 5.4	79.6 ±5.8	73.9 ±5.2	68.2 ±4.5	78.8±4.8	82.7±4.2
Heart rate (b/min)	86.5 ±3.8	98.8 ±4.9[a]	103.7 ± 6.0[a]	99.9 ±3.6[b]	98.9 ±2.7[a]	93.3 ±2.5	84.2 ±7.0[c,e]	80.1 ±4.1[d,e]
Stroke volume (ml)	84.2 ±7.1	81.1 ±9.4	81.7 ± 5.9	73.1 ±8.6	72.7 ±8.8	69.0 ±7.0[b]	87.9 ±8.4	
Cardiac output (l/min)	6.56±1.14	7.56±1.10	7.60± 0.90	6.45±0.97	6.51±0.58	6.31±0.67	7.21±0.84	

[a] $p<0.05$, [b] $p<0.02$ compared to I; [c] $p<0.05$, [d] $p<0.02$ compared to II; [e] $p<0.02$ compared to III

Table 3. Metabolic parameters ($n=10$)

	Examination points							
	I Awake	II Anaesth.	III Surg. +5 min	IV Surg. +25 min	V Surg. +50 min	VI Surg. +75 min	VII Awake	VIII Awake +60 min
$\dot{V}_{O_2}$ (ml/min/m²)		162.9±23.5	150.1±19.5	132.9±9.9[c]	134.0±14.6	126.3±13.2		
Cortisol (µg/dl)	22.1±1.8	20.4± 1.5	17.9± 1.6[b,d]	16.8±1.5[b,d]	16.3± 1.1[b,d]	15.9± 1.2[b,d,e]	15.8±1.4[b,d]	16.1±1.7[a]
Glycaemia (mg%)	79.9±3.2	90.4± 2.7[b]	93.0± 2.2[b]	118.2±6.5[b,d,f]	122.8± 7.3[b,d,f]	139.9±12.7[b,d,f]	133.5±8.9[b,d,f]	145.1±6.7[b,d,f]

[a] $p<0.05$, [b] $p<0.02$ compared to I; [c] $p<0.05$, [d] $p<0.02$ compared to II; [e] $p<0.05$, [f] $p<0.02$ compared to III

Although the constant decrease in cortisol suggests a very good protection against surgical stress during isoflurane anaesthesia, the increases in catecholamines (study still under progress) and in glycaemia might on the contrary suggest too light an anaesthetic plane. Probably the addition of a small dose of analgesic drugs would have abolished this reaction.

Isoflurane for Closed-Circuit Anaesthesia

In a group of 12 patients undergoing minor gynaecological surgery, isoflurane was given in closed circuit. These patients were ventilated with an Engström respirator, specially modified to be completely airtight; the modified closed circuit was connected to an EHN spirometer, allowing the establishment of an O_2/N_2O anaesthesia with injection of liquid anaesthetic into the circuit. Isoflurane was injected by means of a home-made special injection chamber.

The patients received an oxygen/nitrous oxide anaesthesia (mostly 50% O_2/ 50% N_2O). The formula of Lowe was used for injecting liquid isoflurane into the circuit. After a priming dose, the unit dose (according to body weight) was given at predetermined time intervals according to the square root of time (0, 4, 9, 16, 25, 36 min). The actual inhaled isoflurane concentrations were measured by means of an Emma gas analyser or by the Centronic mass spectrometer. The concentrations obtained in the closed circuit are very close to the predictions made by Lowe. The main advantage of this technique is, of course, the economy in consumption of the anaesthetic, as less than 10 ml liquid isoflurane is needed for an anaesthesia of 1 h duration.

The clinical reactions of the patients during this special type of anaesthesia were as might be expected: slight decrease in arterial blood pressure and a slight increase in heart rate.

In conclusion, or clinical experience with isoflurane, together with the data obtained, suggests that the novel inhalation anaesthetic isoflurane is a very promising anaesthetic.

Discussion VIII

Foster: Thank you very much, Prof. Rolly. You used pancuronium. Have you any idea how much potentiation of the muscle relaxation there is with isoflurane?

Rolly: There is a potentiation, but we did not study this.

Siedlecky: Did you measure isoflurane concentration in the first group end-expiratorily?

Rolly: Inspiratorily!

Siedlecky: This does perhaps explain the rise in catecholamines, because with a MAC of 1.0 one could expect to block a rise in the catecholamine levels. I was impressed to see that you had no increase in blood pressure and catecholamines after intubation. How do you block the rise in arterial blood pressure after intubation in hypertensive patients, especially after extracranial vascular surgery?

Rolly: We used isoflurane in a constant concentration from the beginning on. We did nothing special to block a rise in catecholamines. The patient was just anesthetized with etomidate combined with pancuronium and then he was intubated. The increase in catecholamines comes from surgical stimulation.

In some earlier studies we added small doses of alfentanil and in these patients we had much higher rises in catecholamines. We were also astonished to see the remarkable rise in catecholamines after the end of anaesthesia. These patients did not receive any analgesic during the operation but when they felt uncomfortable in the postoperative period they received an analgesic such as Temgesic at once. We were therefore astonished to see such a high rise.

Droh: In Germany we have some difficulty in getting isoflurane. I was lucky to have this drug for 100 patients.

I am very pleased that you do not also promote the method of injecting halothane as Low does.

We have very good vaporizers, and if you are not satisfied with your vaporizers then Dräger must improve them. I think one improvement should be that you can tell from the outside what quantity of anaesthetic has been taken out of the vaporizer. With the Emma from Engström we can measure some volatile anaesthetics such as halothane end-expiratorily. So we are able to say how much anaesthetic has been absorbed by the patient. We can control anaesthesia much better with a vaporizer. We ourselves use normal halothane vaporizers from Dräger and we are very satisfied with them.

I would like to ask one more question. Did you use 50% O_2 and 50% N_2O?

Rolly: Yes.

Droh: I am quite sure that if you go down to 24% O_2 in the inspiratory branch of the closed-circuit system you will not see this rise in catecholamines. I think it is decisive that you go up with the N_2O concentration to over 70%. Then you will have a better analgesic effect of N_2O. We don't like to induce our anaesthesia with etomidate. We usually use barbiturates, either short- or long-acting. Here we also have that "brain-protecting" effect.

Rolly: Well, we used etomidate just to have the pure effect of isoflurane. With respect to your remark about N_2O, I must say that in one of our previous studies we compared N_2O + etomidate with etomidate + alfentanil and we found that with N_2O we had a remarkable sympathetic reaction with a large rise in catecholamines. We did not see this reaction without N_2O, so it was my impression that N_2O caused an increase in catecholamine levels.

Droh: What percentage of N_2O were you using when you saw this increase in catecholamines?

Rolly: Sixty-six percent.

Droh: I would like to ask you to repeat your study with a N_2O percentage of over 70%. I cannot believe that N_2O would have such an effect.

McIntyre: I would like to make two brief comments. First with reference to the rise in pulse rate. We found that once in a while, especially during abdominal surgery, even if huge quantities of isoflurane were given we saw a tachycardia. In these cases we avoid pancuronium and give morphine or fentanyl intravenously. Sometimes we use propanolol to reduce the pulse rate.

The second comment is with reference to what Dr. Rolly said about the metabolism of isoflurane. There is so far insufficient evidence in the literature to reassure us that we can use this drug for lengthy surgery in obese patients, particularly in patients who have renal problems. I think there is insufficient evidence in the literature that this drug is really safe. It may be better than halothane or Ethrane, but this is still an open question.

Quantitative Studies of Oxygen Exchange in Blood Substitution by Polymeric Stroma-free Haemoglobulin Solution

G. Klein, U. Ottermann, H. Förster, and R. Dudziak

Summary

Quantitative studies of O_2 exchange in blood substitution through polymeric stroma-free haemoglobulin solution show that the O_2 supply to the tissue is not reduced and central venous PaO_2 remains in the normal range. Sixty per cent of the O_2 is carried in the plasma phase and 40% is carried by the erythrocytes. It is concluded that in the normal pO_2 range this stroma-free haemoglobulin solution can be used as O_2-carrying medium.

Experiments with Haemoglobin Solutions as Perfusion and Infusion Solution

U. Ottermann, S. Klein, K. Bonnhard, and R. Dudziak

Summary

Stroma-free haemoglobulin solutions have the same O_2 capacity as the haemoglobulin of the erythrocytes. In animal experiments such haemoglobulin solutions were able to maintain a sufficient tissue oxygenation after massive blood loss. The dogs even survived extreme isovolumetric haemodilution without any damage to the tissue.

The Technical and Physiological Aspects of Surgery and Anaesthesia Without Homologous Blood

R. Lapin, W. Erdmann, and N. S. Faithfull

Introduction

Performing major surgical procedures without the use of homologous blood transfusions is an idea whose time has come, especially with blood supplies becoming more limited due to the reluctance of donors who are, in general, frightened of the AIDS epidemic.

Our experience with 6000 surgical procedures performed without utilizing blood transfusions in the past 10 years has taught us that patients can indeed be operated on safely without utilization of banked blood.

In approaching this problem, we began by making a few assumptions which the years have proven to be reasonably correct.

1. The surgical patient must serve as his own blood bank.
2. If a patient must be transfused, then the best possible blood for such a patient is his own.
3. The way to reduce or eliminate transfusions in elective surgery is to avoid its loss by improving techniques and instrumentation.
4. Giving a patient a homologous blood transfusion is analogus to giving a "liquid organ transplant."
5. Surgery and anaesthesia are unphysiological states and as such should be reduced to a absolute minimum.
6. The magic number of 10 haemoglobin below which anaesthesia is generally considered unsafe is without foundation.

In implementing the above points in an active surgical practice which covers all surgical specialties, we have the following plan:

1. All surgical patients are carefully and rapidly diagnosed and followed until discharge by internal medicine specialists.
2. Hypotensive anaesthesia is used whenever possible in the appropriate patient in order to reduce blood loss and operative time.
3. All surgeries are carried out with maximum intraoperative monitoring, which is continued in the intensive care unit.
4. Electrocautery is the preferred mode of operating because of the great reduction in operative time and blood loss.
5. All significant blood lost at the time of surgery is returned to the patient by an autotransfusion device.
6. Volume replacement is carried out using crystalloids and hydroxyethyl starch.

7. In significantly anaemic patients needing increased oxygen carrying capability, perfluorochemicals (fluosol-DA 20%) are employed preoperatively. Anaesthesia in these patients is carried out by employing ketamine, diazepam, and muscle relaxants.
8. Total dose intravenous iron dextran supplemented by folic acic, B-12 and decadurobulin is given to the severely anaemic patients.

Using the above plan we were able to approach all major cases with which we were presented with excellent results.

Fluid Replacement

Sudden loss of more than 10%–15% of the circulating blood causes a corresponding fall in filling pressure and a marked reduction in minute volume. In order to overcome this fall in the minute volume, catecholamines are released which cause a rise in vascular tone. In spite of the blood volume reduction by blood loss, filling pressure and minute volume are maintained by compensatory mechanisms as long as the volume lost is not enough to exceed the autoregulative capacity. Even with a blood loss of 10%–15%, full compensation is not achieved without replenishing the blood volume unless the fluid lost from the vascular bed is less than 10% of the total volume. This loss of volume can be easily and safely replaced with crystalloid solutions.

Haemoglobin is also lost with the blood, and it is reasonable to replenish the intravascular fluid level with whole blood and therefore not only restore the minute volume but also the oxygen transport capacity. However, blood availability is limited for the following reasons:

1. The number of donors is limited.
2. Blood only keeps for three weeks.
3. Whole blood can only be stored in local banks and transported at high costs at a temperature of 4°–6 °C, with critical shortages in the event of a disaster with mass casualties.
4. The incidence of infection transmitted by donor blood has not been abolished.
5. Respect for the religious convication of patients rejecting transfusion of blood or blood derivatives presents a moral and ethical dilemma to the majority of doctors.
6. The still unanswered problems of the long term immunological effect of blood transfusions on the recipient which has recently been brought into focus by the outbreak of the AIDS epidemic.

It is therefore only logical to try and break the dependency on homologous blood transfusions whenever and wherever possible and look for other alternatives.

Colloidal Blood Substitution and Its Limitations

Arterial O_2 transport capacity is 1000 ml O_2/min at a cardiac minute volume of 5/min ([500 ml × 0.95 (% Hb-saturation) × 0.15 (g/ml Hb) × 1.39 (ml/g O_2-binding capacity or Hb)] + [5000 ml × 0.003 (ml O_2 dissolved in plasma)]). Only 300 ml O_2 are consumed per minute. Taking into account that the critical venous value of 19 mm Hg PO_2 corresponds to 34% Hb-saturation and that an excessive decrease in haemoglobin concentration is primarily compensated for by an increased cardiac output, then the haemoglobin concentration may fall to 50% normal (hct 22%) without impairment of the tissue's oxygenation. The critial value is only reached with a hematocrit of 15%. Thus, there exists a broad safety margin for tissue oxygen supply, provided that the myocardium remains healthy and the filling pressure, dependent on the quantity of intravascular fluid, does not decrease or is immediately restored. Therefore, the fall in O_2 transport capacity from haemoglobin loss can be largely compensated for by merely replenishing the vascular system.

Among the numerous colloidal blood substitutes, dextran, and gelatin derivatives have until now met the following criteria:

1. Osmotic pressure corresponds to that of plasma.
2. Viscosity is not higher than that of plasma to allow capillary perfusion.
3. It is free of toxic or pyrogenic material.
4. It must stay in the vascular system for sufficient periods.
5. It must not interfere with coagulation.
6. It must not be stored in the body.
7. It must be sterilizable.
8. It must be capable of being given without time-wasting preparation and should be kept at normal temperatures.
9. It must not be carcinogenic.

Blood substitutes are 10 times cheaper than protein preparations, a great economic advantage.

Physicians differ as to how to replace blood volume in shock. They believe replenishing extracellular fluid is more important than the intravascular fluid. Their recommendation to administer electrolyte solution (e.g., Ringer's lactate instead of colloids, up to the limit of the hematocrit value) has certain grave pitfalls: the interstitual space (15% of body weight) has twice the size of the vascular space (8% of body weight), and thus only one-third of the fluid introduced stays within the vascular bed and three times more solution has to be given than colloid to achieve the same effect. With the excessive electrolyte infusions needed, the considerable increase in intestitial water content may lead to interstitial edema, especially in the lungs, and in the presence of other aggravating factors, shock lung may readily develop.

The replacement of blood loss by colloidal non-oxygen-carrying blood substitutes is limited by the lack of oxygen transport capacity (0.3 ml O_2/100 mm Hg PO_2 in 100 ml) as compared to blood with 15 g Hb; this $15 \times 1.39 = 20.58$ ml in 100 ml. Experiments with stroma-free haemoglobin solutions to replace blood failed because of high kidney toxicity and rapid elimination. The critical considerations of synthetic oxygen-binding chelates so far have not reached the stage of animal experimentation, but oxygen-carrying fluorocarbons have reached the stage of clinical investigation.

Perfluorochemicals

In 1966, Clark and Gollan described an experiment in which a mouse could be kept for several hours in a glass filled with an organic perfluorocarbon solution, breathing the oxygenated fluid without obvious impairment to oxygen tension. In 1968, Geyer et al. successfully replaced all of the blood in rats with a perfluorocarbon emulsion without any apparent ill effect in these totally blood-free animals. Clark succeeded in 1970 in doing the same in larger animals such as dogs.

Meanwhile, active research was launched and several international congresses have gathered the results in their proceedings.

Following Clark's demonstration of the feasibility of perfluorochemicals as oxygen carriers, more than 30 compounds have been tested for their characteristics from which just 6 have shown to be usable thus far.

The compounds of interest to our topic are all completely fluorinated derivatives of hydrocarbons and related compounds (+ perfluorocompounds lacking any hydrogen atom, other halogen atom, or double bond). Due to the strong fluorobonds with the carbon, these molecules are very stable, inert, not polarizable and unreactive as the morecule is excellently protected by a uniform shield of fluorine atoms.

Clark and Geyer found that particles larger than 0.6 µm tend to produce microagregates occluding the capillary mesh. This is overcome by emulsification through either ultrasonic treatment or high-pressure emulsification by means of high shear rates, the emulsified preparation is stabilized by egg yolk phospholipids (lecithin) and a polyoxyethylene-polyixypropylene polymer (Pluronic F68) which had already been used in the early sixties in the production of fat emulsions for parenteral nutrition and also has oncotic effects.

The perfluorocarbons are excreted by vaporization via the lungs and partially by transpiration. Thus, their excretion rate is related to vapor pressure (Table 1).

Perfluorocarbons (PFC) as artificial blood substitutes are used in solutions (Table 2) containing 20% and 35% PFC whereby the 20% solution has a lower viscosity than blood. To the basic electrolyte solution, hydroxyethyl starch is

Table 1. Advantages and disadvantages of PFC compounds in relation to boiling point and vapor pressure

	Boiling point (°C)	Vapor pressure	Advantages and disadvantages
Perfluorobutyltetrahydrofuran (FC-75)	102	58	Due to high vapor pressure (in the lungs) it causes pulmonary edema, is short acting and is therefore unsuitable.
Perfluorotripropylamine (FTPA)	130	20	No lung problems; despite too rapid elimination, can be defined as good.
Perfluorodecalin (FDC)	142	12.7	Not too rapidly excreted no lung complications, no longterm body storage, excellent characteristics.
Perfluorotributylamine (FC-43)	176	1.1	Very slow elimination (months), long-term storage in the body (liver, spleen, lungs). Overall characteristics good.

Table 2. Composition of Fluosol

	Fluosol	
	−DA 20%	−DA 35%
Perfluoro-		
tripropylamine	6.0	10.5
decalin	14.0	24.5
Pluronic F68 (%)	2.7	2.7
Egg yolk phosphatide	0.4	0.4
Glycerol	0.8	0.8
Hydroxyethyl starch	3	3
Glucose (mmol/l)	10	9.1
Na^+	128	117
K^+	4.6	4.2
$Mg\,2^+$	2.1	1.9
$Ca\,2^+$	2.5	2.3
Cl^-	112	102.5
$HCO_3{}_-$	25	23
Osmotic pressure (mosm/l)	410 (320)	
Oncotic pressure (mmH$_2$O)	380–395	

added to achieve the osmotic pressure of blood and sodium bicarbonate as a buffer, otherwise CO_2 transport is impaired. A combination of rapidly eliminated FDC with the longer lasting FC-43 has been shown to be ideal. The amount of oxygen transported by fluorocarbons is proportional to the partial pressure, which means that hyperoxygenation (breathing 70%–100% oxygen) is advisable. Taking into account that the oxygen partial pressure at the oxygen delivery capacity of a 2% PFC solution at 100% FiO_2 (about 550 mm Hg PaO_2) amounts to 5.0 ml/min, corresponding to the capacity of blood with 40% Hct and being far higher than that of 3% Hct (Fig. 1); a 35% fluorocarbon solution has a decisively higher delivery capacity, but the higher viscosity reduces the capillary perfusion ratio at the same time. Experiments have shown that the oxygen delivery capacity

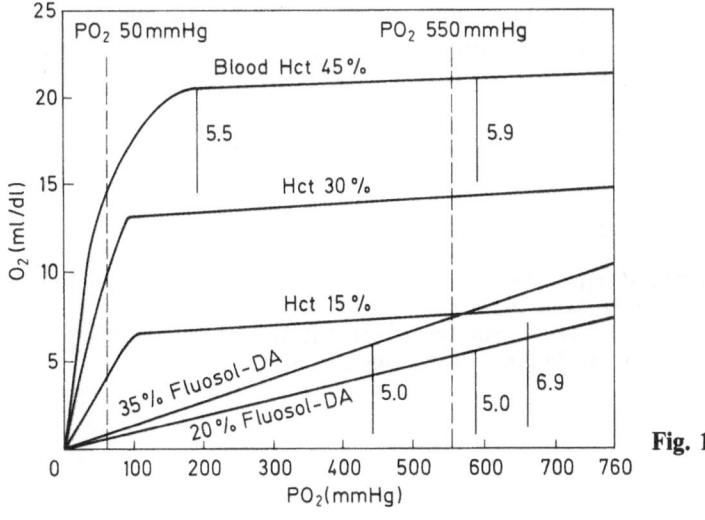

Fig. 1

121

of PFC in a 35% solution while breathing 50% oxygen is still far better than that of blood with 15% Hct.

The combination of rapidly eliminated FDC with rather slowly eliminated FC-43 keeps oxygen transport capacity dependent on perfluorocarbon long enough at a suitable level to allow the erythrocyte count to increase to a sufficient level to take over this function again. The rather prolonged effect of hydroxyethyl starch (HES) keeps the colloid osmotic pressure up until resynthesization of plasma proteins has occurred.

Perfluorocarbons have been extensively tested in animal experiments, and human trials in Japan, Germany, Austria, Canada, Holland, and the United States have shown no significant negative side effects. In the United States, fluosol is used as a blood substitute in a controlled multicentre study under guidelines set by the Food and Drug Administration. The study is primarily limited to Jehovah's Witness patients. The preliminary data has been encouraging.

Its small particle size as well as its ability to carry oxygen has given researchers a new tool for the study and treatment of microcirculatory disorders. Experimental as well as clinical work is now being conducted in the treatment of myocardial infarctions and cerebrovascular accidents using fluosol. Other fields where PFCs have shown great promise are organ preservation and transplantation, radiation therapy, shock, and radiology.

Fluosol meets all the demands made on colloid blood substitutes as well as the ability to carry oxygen, carbon dioxide and other gases. The only draw-back is that high O_2 levels are needed to make the oxygen transport potentials comparable to that of normal blood. They then represent a good substitute for blood and erythrocytes without blood group incompatibilities or other reactions in the recipient; the danger of transmission of hepatitis, AIDS, or other infection is avoided; it can be made available in large quantities, and it is compatible with many drugs, nutrients, etc. At present, freezing is still required, but in the second generation of fluosol, this need will be eliminated. Cost effectiveness and competition with homologous blood will make this product cheaper, safe, and available in large quantities for the treatment of all mankind.

In conclusion, we can state that bloodless surgery is indeed a reality and not a myth and that our experience supports this statement. We can only hope that other physicians will follow a similar course which will add more strength and credibility to the notion that blood is potentially a dangerous drug and its use in medicine should not be taken lightly.

References

Clark LC (1970) Whole animal perfusion with fluorocarbon dispersions. Federation Proceeding 29:1695

Clark LC, Gollan F (1966) Survival of mammals breathing organic liquids equilibrated with oxygen at atmospheric pressure. Science 152:1755–1756

Geyer RP, Monroe RG, Taylor K (1968) Survival of rats having red cells totally replaced with emulsified fluorocarbon. Federation Proceeding 27:384

The Ability of Perfluorochemicals to Provide Microcirculatory Oxygenation

N. S. Faithfull, A. R. Smith, W. A. van Alphen, M. Fenema, W. Erdmann, R. Lapin, C. E. Essed, and A. Trouwborst

Introduction

Perfluorochemicals have recently become available in the form of oxygen-carrying plasma substitutes containing 20% or 35% (weight by volume) of perfluorocarbons respectively. These preparations are known as Fluosol-DA 20% and Fluosol-DA 35%. The composition of these products is shown in Table 1. They have a number of interesting properties in relation to their oxygen-carrying ability and their ability to penetrate the microcirculation. This paper will discuss some of these properties.

Table 1. Composition of Fluosol-DA 20% and 35% emulsions (after Naito and Yokoyama 1978)

	Fluosol		
	$-DA\ 20\%$		$-DA\ 35\%$
Perfluoro-tripropylamine (FTPA)	6.0		10.5
decalin (FDC)	14.0		24.5
Pluronic F-68 (%)		2.7	
Egg yolk phosphatide		0.4	
Glycerol		0.8	
HES		3.0	
Glucose (mmol/l)	10		9.1
Na^+	128		117
K^+	4.6		4.2
Mg^{2+}	2.1		1.9
Ca^{2+}	2.5		2.3
Cl	112		102.5
HCO_3	25		23
HPO_4			
SO_4			
Osmotic pressure (mosm/l)		410 (320)	
Oncotic pressure (mmH$_2$O)		380–395	

123

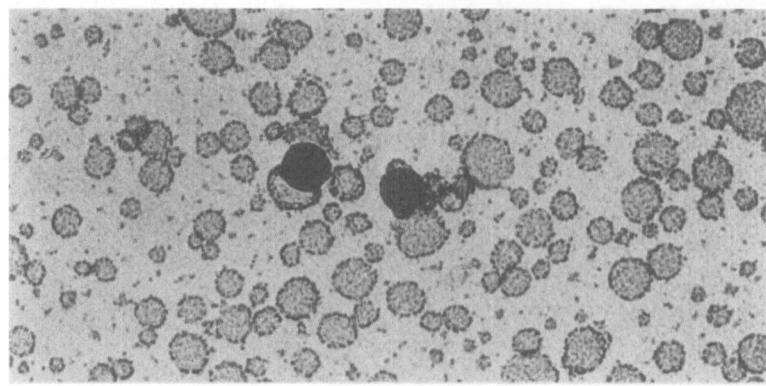

Fig. 1

Particle Size of the Emulsion

Though pure fluorocarbons at 37 °C can contain 40 or more volumes per cent of oxygen they are immiscible with blood, and hence, in order to prevent embolic phenomena, they are supplied in an emulsified form. The particle size of the perfluorocarbons varies between 0.6 μm and less than 1 μm, with more than 90% being in the range of less than 0.2 μm diameter (Yokoyama et al. 1974). Figure 1 is an electron micograph of the emulsion containing two polystyrene latex granules of 0.1 μm in diameter. It can be seen that the emulsion particle size is very small in comparison to an erythrocyte, which has an average diameter of 7.2 μm and a thickness of 2.2 μm at the rim, decreasing to 1 μm in the centre (Price-Jones 1933).

The small particle size ensures that a very large surface area is available, both for uptake and release of oxygen. The surface/volume ratio of the perfluorochemicals present in Fluosol emulsions is very high, and ensures rapid gas exchange by facilitating both uptake of oxygen in the pulmonary capillaries and release of oxygen in the tissues.

Oxygen Carriage and Release

The oxygen content of fluorocarbon emulsions is dependent on the partial pressure of oxygen and, unlike haemoglobin, the amount of oxygen that can be taken up is directly proportional to the oxygen tension. Due to the S-shaped oxyhaemoglobin dissociation curve, this is not so in the case of haemoglobin, and uptake and release of oxygen take place over a very narrow physiological range. Indeed, normal adult haemoglobin changes from 20% oxygen saturation to 80% saturation when the partial pressure is changed by only a matter of 30 mm Hg.

At full saturation, normal blood will contain something in the region of 20 ml oxygen per 100 ml blood and this will be achieved at a PO_2 of about 120 mm Hg. At this PO_2 Fluosol-DA 20% will contain approximately 1.2 ml oxygen per 100 ml emulsion. This amount, though small, is nevertheless about three times as much as can be carried by plasma alone.

The oxygen content of Fluosol is directly proportional to the partial pressure of oxygen and, at 550 mm Hg, the content will still only be 5.5 ml per 100 ml. The content of blood will have also increased slightly at this PO_2 due to the increase in dissolved oxygen, and will have risen by about 1.3 ml per 100 ml. It can be cal-

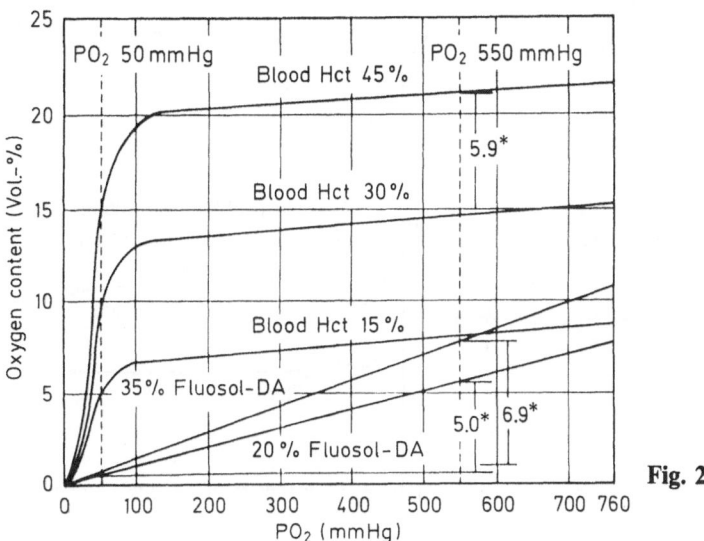

Fig. 2

culated that Fluosol-DA 20% will not contain as much oxygen as normal blood, unless the PO_2 is raised to approximately 3000 mm Hg – at which point, even if obtainable, the danger of oxygen toxicity will be of overwhelming importance.

Fortunately, the oxygen content of Fluosol is not of great importance. What is of importance is its ability to deliver oxygen to the tissues. Figure 2 shows the oxygen content of Fluosol-DA 20% and 35%, and blood at various haematocrits as a function of the PO_2. The straight line "dissociation curve" of Fluosol is immediately apparent. This is to be compared with the S-shaped relationship between blood oxygen content and PO_2.

The vertical dotted line at a PO_2 of 50 mm Hg represents tissue PO_2. It should be stressed that it is almost impossible, in fact, to quantify tissue PO_2, which is of course influenced by the supply and demand situations in each organ. It may also vary widely over small areas in the same tissue (Erdmann and Vogel 1973; Faithfull et al. 1983a). However, it helps in our understanding of oxygen release phenomena to take one "mean value" of tissue PO_2. As can be seen from Fig. 2, the release of oxygen from the 20% preparation of Fluosol when the PO_2 is decreased from 550 mm Hg to 50 mm Hg amounts to 5.0 ml. This is very little less than the 5.9 ml released by blood with a haematocrit of 45% over the same range, and considerably more than that released by blood with a haematocrit of 30%.

Viscosity

Resistance to flow along a tube (or blood-vessel) depends on a number of factors, and the sum of these factors is known as the "vascular resistance" of a system. This is usually expressed as a pressure/volume relationship – in other words the amount of pressure necessary to produce a certain flow in the system. Total resistance to flow is made up of the product of a vascular factor, determined by the physical characteristics of the vessels (diameter, branching etc.), and the viscosity of the liquid (blood) that is to be forced through. In other words, if the viscosity is halved the resistance to flow is halved.

125

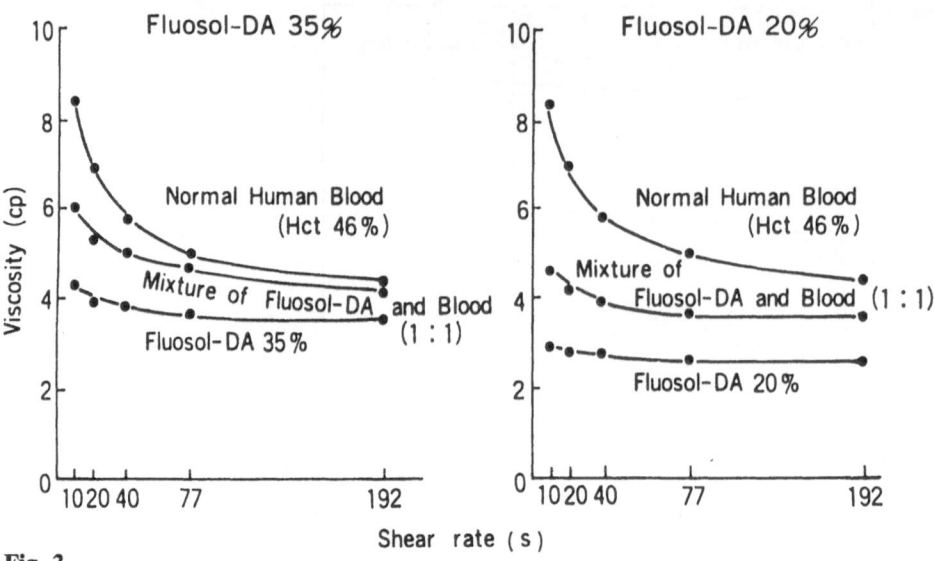

Fig. 3

Blood is a non-Newtonian fluid; in other words it is a liquid whose viscosity changes with shear rate (or flow rate), and at low flows its viscosity can be 10–100 times that at high flows. This situation will, of course, tend to impede flow in the microcirculation.

Two important factors tend to improve the circulation through capillaries. In the first place, the phenomenon of "plasma skimming" takes place, in which the smaller vessels in the microcirculation receive proportionately more plasma and fewer erythrocytes than are present in the general circulation. In other words, the haematocrit is decreased and hence viscosity is decreased in the smallest vessels.

A further factor that improves microcirculation is the so-called Fahraeus-Lind-quist phenomenon, whereby erythrocytes can flex and deform to such an extent that they can pass through vessels of considerably smaller diameter than themselves. This is so effective that the blood then behaves like a Newtonian fluid with a viscosity that does not differ from that of pure plasma (Goslinga 1982). This phenomenon only holds good to a certain extent and, at critical diameters of the capillaries, the so-called inversion phenomenon occurs, whereby further decrease in capillary diameter results in a sudden sharp increase in viscosity. Under certain conditions the inversion phenomenon occurs at a greater vascular diameter than normal. These conditions include acidosis, in which vascular flexibility is decreased (Schmid-Schönbein et al. 1973), and platelet aggregation.

The viscosities of Fluosol-DA 20% and 35% are shown in the graphs in Fig. 3. Also shown are the viscosity curve of normal blood with a haematocrit of 46% and the viscosity of a one-to-one mixture of Fluosol and blood. It can be seen that though blood is approximately 1.8 times as viscose as Fluosol-DA 20% at the highest shear rates, it is about 2.9 times as viscose at low rates. In practice, blood is usually mixed with Fluosol and it can be seen that, at low shear rates, full blood is 1.9 times as viscose as a 50% mixture of blood and Fluosol.

The above results (Naito and Yokoyama 1978) have considerable relevance to the microcirculation and it can be argued that haemodilution with Fluosol will assist microcirculatory flow. This should be particularly so under conditions of sludging or hypoxic acidosis, such as may occur in conditions of infarctive isch-aemia. It should also be noted that haemodilution with Fluosol-DA 20% will in-crease total body oxygen transport without increasing cardiac work. This is due to the reduction of peripheral resistance and consequent increase in cardiac out-put. These effects on oxygen transport occur even when room air is being breathed (Faithfull et al. 1983 a).

Experimental and Clinical Investigations

Experimental work based on the above properties of Fluosol-DA 20% has been carried out in the Erasmus University, Rotterdam, and tends to support strongly the conclusion that Fluosol can penetrate into areas of impaired microcirculation. Due to its low viscosity, it can traverse long collateral perfusion pathways, and its small particle size enables impacted erythrocytes to be bypassed and reoxygen-ated. Hence the vicious circle of sludged standstill can be overcome. These studies are reported in greater detail elsewhere (Faithfull et al. 1983 c).

Myocardial Studies

In anaesthetized pigs the chest was opened and four oxygen microelectrodes were implanted to a depth of 4 mm in the myocardium supplied by a chosen branch of the left anterior descending coronary artery (LAD). A schematic diagram of the experimental set-up is given in Fig. 4.

After occlusion of the chosen branch of the LAD, rapid falls of PO_2 occurred and no treatment resulted in further slow falls of PO_2 in the ischaemic areas over the next 4 h. Bleeding (20 ml/kg body wt.), followed by reinfusion and lowering of blood viscosity with non-oxygen-containing dextran solutions, markedly de-creased the oxygenation state still further.

Haemodilution using Fluosol-DA 20% caused an increase of more than 100% in postocclusion myocardial PO_2 values, which continued to rise slowly during the whole experiment. Though cardiac output was increased by 35%, overall car-diac work was not increased due to decreased viscosity produced by the fluoro-

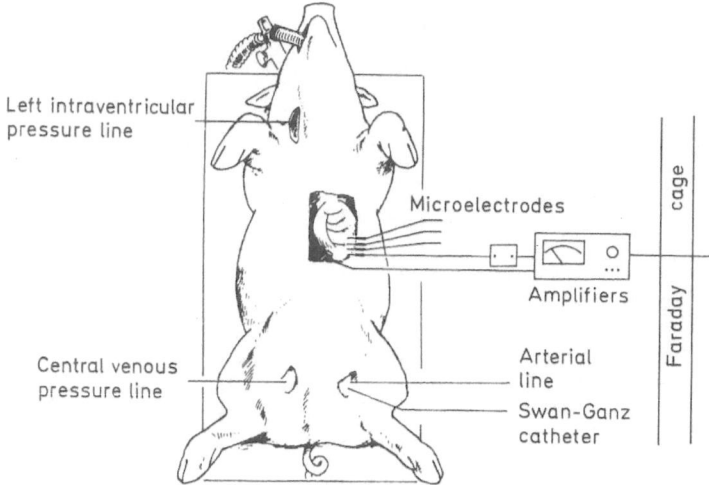

Fig. 4

carbon solution. Decrease in both size and severity of the resulting infarct areas was confirmed by histology, confluent waviness of myocardial fibres being taken as an early sign of myocardial ischaemic damage (Bouchardy and Manyo 1974).

Perfusion of Amputated Extremities

A number of traumatically amputed human extremities which were unsuitable for immediate replantation were perfused with Fluosol-DA 20%. A schematic diagram of the experimental set-up is shown in Fig. 5. The main supplying artery of the extremity or digit was cannulated and perfusion commenced as soon as possible. The perfusate (Fluosol-DA 20%) was cooled to 5 °C and oxygenated by a bubble oxygenator using a gas mixture of 95% O_2 and 5% CO_2. It was circulated by pulsatile pump and the mean arterial pressure was kept at about 90 mm Hg as measured in the in-going perfusion line by an aneroid pressure gauge.

A typical recording of the PO_2 measured by the oxygen electrodes is shown in Fig. 6. This was measured in a thumb which had suffered 1 h of warm ischaemia and 6 h of cold ischaemia. As can be seen, within half an hour after commencement of perfusion, tissue PO_2 levels in the digit began to rise. There was usually an "overshoot" of PO_2, which then fell to lower levels. This would indicate hyperaemia followed by vasoconstriction and indicates viability of the extremity. At the position of the electrode indicated by the continuous line, the PO_2 rose only after a delay of about 1 h and then rose very sharply. The overshoot was small

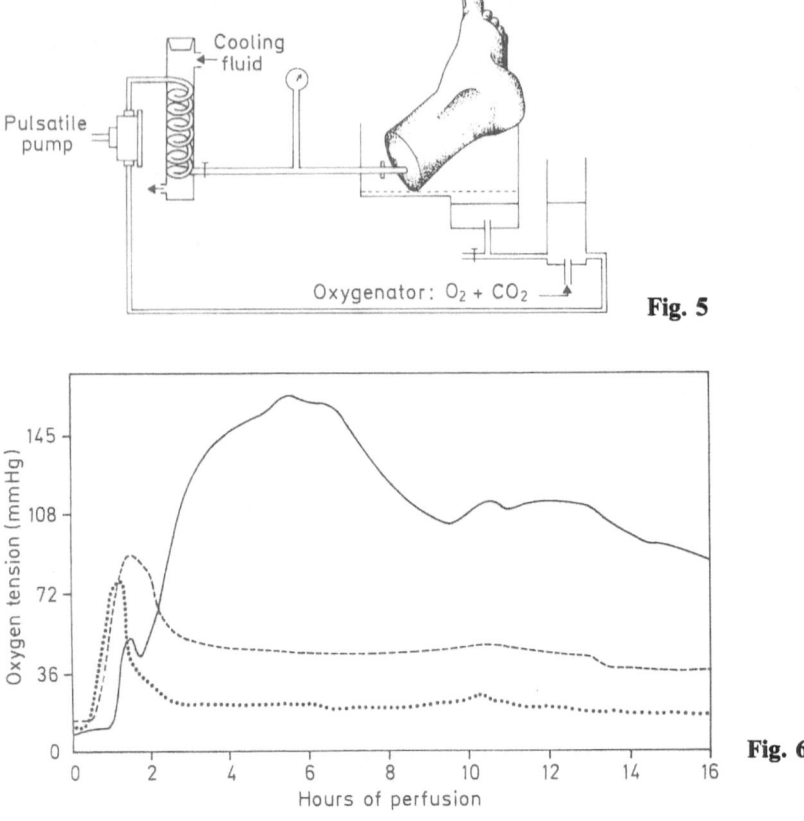

Fig. 5

Fig. 6

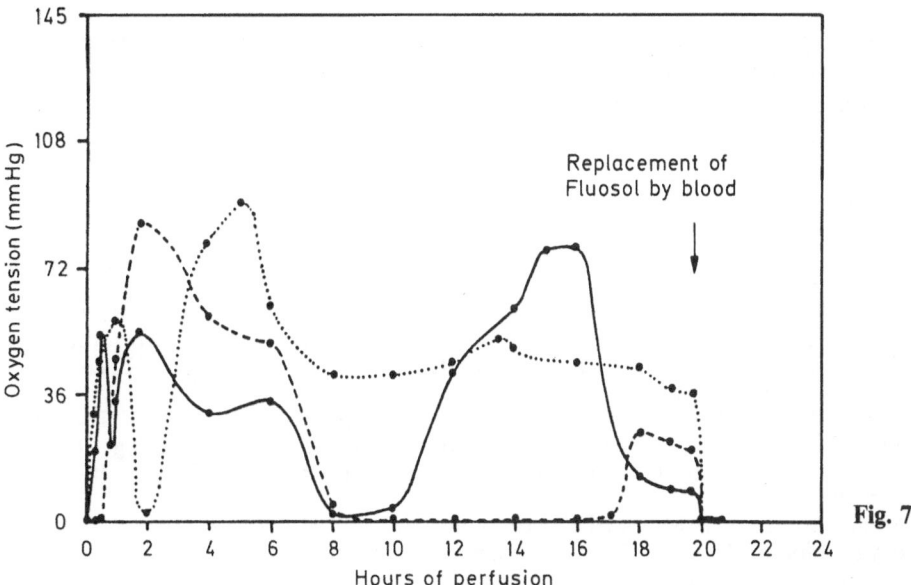

Fig. 7

and was followed by a sustained rise. These changes may be interpreted as indicating that the electrode was in the vicinity of a blood-vessel, the perfusion of which was delayed – possibly due to obstruction by "stiffened" erythrocytes.

The ability of Fluosol to penetrate the microcirculation is well demonstrated in Fig. 7. After 20 h of perfusion of this finger, at which time all three electrodes were indicating reasonable levels of PO_2, the perfusate was changed from Fluosol to blood that had been cross-matched and found to be compatible with the patient in question. As can be seen, measured PO_2 values fell rapidly to zero and remained at that level. It may well be argued that failure to perfuse with blood was caused by the oedema formation which was always seen to a certain degree during perfusion, in spite of the hyperosmolar character of the perfusate. This oedema was reflected by weight-gain of the extremities, which usually amounted to about 20% after 24–48 h of perfusion. This would tend to compress the microcirculation from without and impede circulation. Figure 7 gives a graphic demonstration of the better penetrability of Fluosol in comparison with blood under these circumstances. To date two successful implantations have been performed following this technique of perfusion (Smith et al. 1983).

General Conclusions

The above work has demonstrated certain important characteristics of fluorocarbons with respect to tissue oxygenation. In the first instance, it has been demonstrated that extracorporeal perfusion with Fluosol-DA 20% can maintain oxygenation and viability of isolated extremities for periods of up to 48 h.

The ability of Fluosol to penetrate compromised capillary beds has been demonstrated. Its low viscosity enables it to travel along collateral perfusion pathways and to provide oxygenation of areas of ischaemic hypoxia. The low particle size may, in combination with its low viscosity, contribute to the re-establishment of perfusion in areas of circulatory arrest.

The clinical implications of the above findings are numerous and widespread. The importance of Fluosol in maintaining and improving microcirculatory oxygenation should not be underestimated.

References

Bouchardy B, Manjo G (1974) Histopathology of early myocardial infarcts. A new approach. Am J Pathol 74:301

Erdmann W, Vogel HR (1973) Die Sauerstoffversorgung des Gehirns. Bundesärzteblatt 8

Faithfull NS, Erdmann W, Fennema M (1983a) Oxygen supply to the myocardium. In: Proceedings of the International Society for Oxygen Transport to Tissue. Rushton, Los Angeles 16–19 August 1983

Faithfull NS, Fennema M, Erdmann W, Lapin R, Smith AR, Van Alphen W, Essed CE, Trouwborst A (1983b) Tissue oxygenation by fluorocarbons. In: Proceedings of the International Society for Oxygen Transport to Tissue. Rushton, Los Angeles 16–19 August 1983

Faithfull NS, Fennema M, Essed CE, Erdmann W, Jeekel H, Lapin R (1983c) Collateral oxygenation of the ischemic myocardium: The effect of viscosity and oxygen carrying fluorocarbons. In: Advances in blood substitute research. Liss New York, p 229

Goslinga H (1982) The viscosity of blood. An experimental study into the effects of alterations in blood viscosity during shock. PhD Thesis, University of Utrecht. Drukkerij Elinkwijk, Utrecht

Naito R, Yokoyama K (1978) Perfluorochemical blood substitutes. Green Cross, Osaka

Pryce-Jones C (1933) Blood pictures: an introduction to clinical haematology. Williams and Wilkins, Baltimore

Schmid-Schönbein H, Weiss J, Ludwig H (1973) A simple method for measuring red cell deformability in models of the microcirculation. Blut 16:369

Yokoyama K, Suzuki A, Utsumi I, Naito R (1974) Determination of particle size distribution fluorocarbon emulsion by means of centrifugal sedimentation. Chem Pharm Bull (Tokyo) 22:2966

Discussion IX

Bonnhard: Did you bubble-oxygenize the dextran solution? If so, then the solubility of O_2 in this aqueous solution is only one-third of the solubility of O_2 in fluorocarbon.

Faithfull: No, this was not a bubble-oxygenated solution. The blood of the animal was replaced by fluorocarbon in similar volumes and the same happened with dextran. The animal itself was, of course, being oxygenated. This was not an isolated heart-lung preparation, but the whole animal.

Siedlecky: Johanson from Copenhagen reduced haemoglobin to 4% and replaced it with dextran. He ventilated his animals with 100% oxygen and he was successful in keeping his animals alive. Could you please comment on this?

Lapin: A haemoglobin of 4% is, of course, quite compatible with life, regardless of what you replace the volume with. The main idea of using Fluosol is to supercharge oxygen in the plasma phase. As you know, 80% of the utilizable oxygen is extracted from the plasma before the red cells are touched.

And therefore if you have a patient in a hyperbaric chamber you don't need any red cells to maintain his life. The studies of Johanson are not really comparable. Dextran will not carry any oxygen. The margin of safety with respect to haemoglobin concentration is about 5%, as Prof. Erdmann showed in his studies. But essentially all you need is 300 ml circulating per minute in order to sustain life.

Droh: Is it possible to increase the coloid osmotic pressure with Fluosol?

Faithfull: No, Fluosol is iso-oncotic.

Rieger: Dr. Faithfull, you measured the haemodynamics of the heart. Did you also measure the microcirculation?

Faithfull: No, we did not measure the microcirculation, but we measured oxygenation of the tissue directly.

Analytical Aspects of Some Anaesthesiologically Important Peptide Hormones

H.-P. Kamin

Introduction

Radio-immunoassays (RIA) are extremely sensitive methods for the radio-immunochemical determination of hormones in biological fluids. However, the need for alternative immunochemical techniques continues and has led to the development of, e.g., fluorescent immunoassays (FIA) (Ullmann et al. 1976; Borner 1981), enzyme immunoassays (EIA) (Van Weemen et al. 1971; Rubenstein et al. 1972), and zone immunoelectrophoresis assays (ZIA) (Vesterberg 1980). Electrophoretic work in this field depends on the interaction of the hormone with a precipitating agent and the visualization of the precipitate. The present work deals with the observation that in the case of β-endorphin and arginine-8-vasopressin in agarose gels containing antisera, special buffers and an optical brightener, it is possible to develop peaks and to visualize them by ultraviolet radiation using one- and two-dimensional techniques.

Materials and Methods

For gel preparation standard electroendosmosis agarose ($M_r = -0.13$) from Bio-Rad, Richmond, California, was used. Guinea-pig antilysine vasopressin serum and rabbit antihuman β-endorphin serum came from Immuno Nuclear, Stillwater, Minnesota.

The buffer solutions used below were composed of citric acid 1-hydrate (G.R.) and disodium hydrogen phosphate 2-hydrate (G.R.) (Merck, Darmstadt). Pure human β-endorphin, pure arginine-8-vasopressin and bovine serum albumin (BSA) were purchased from Serva, Heidelberg. BSA was reconstituted with deionized water to give a solution of 50 mg/ml.

One-Dimension Procedure

β-Endorphin and vasopressin were dissolved in deionized water to give antigen solutions of 0.5 mg/ml and 1.0 mg/ml respectively. One per cent (w/v) solutions of agarose in 0.01 M citrate–0.02 M phosphate buffers (pH = 4, 5, 6, and 7) after McIllvaine (Ohlenschleger et al. 1980) were prepared. After these had been cooled down to 40 °C, an optical brightener (10 μl/ml gel) and the antiserum (5 μl/ml gel) were added to each solution. The mixtures were poured on to microscopic slides to give gel layers 1.5 mm in thickness. Sample wells, 3 mm in diameter (one in each gel), were punched and the holes were filled with 5 μl of the corresponding antigen solution. Electrophoresis was carried out at 10 °C on a Bio-Rad horizontal electrophoresis cell, model 1415, applying a voltage of 20 V/cm for 30 min.

Two-Dimensional Procedure

A 70-mm × 15-mm bed of agarose gel was prepared on one side of a 70 mm × 70 mm glass plate, the remaining area of 70 mm × 55 mm being covered with another glass plate. Buffers with pH values 4 and 5 and the gel matrix were composed as mentioned above. A sample well was cut on each plate and filled with a mixture of 4 μl antigen solution and 4 μl BSA solution. The first-dimension electrophoretic run was performed under the same conditions as above. After a separation time of 30 min the covering glass plate was replaced by a 1% agarose gel containing buffer, optical brightener and antiserum in the amounts mentioned above. A voltage of 20 V/cm was applied at right angles to the first-dimension pattern. Electrophoresis was running at 10 °C for 30 min. Anodic and cathodic buffer reservoirs were connected with the plates by ultrawicks (Bio-Rad) according to the isoelectric points of the hormones [9.9 for β-endorphin (Santagostino et al. 1982) and 10.9 for arginine-8-vasopressin (Decker et al. 1968)].

Visualization

The developed peaks were made visible by means of a 366-nm ultraviolet source. This was followed by the standard dyeing procedure using Coomassie brilliant blue R 250.

Results

The electrophoretic experiments outlined above were carried out in close relation to Laurell's rocket immunoelectrophoresis (Laurell 1966) and the crossed immunoelectrophoresis of Clarke and Freeman (1968), but quantitative aspects were not stressed. In the one-dimensional procedures, β-endorphin showed peaks at ph values of 4 and 5, and vasopressin at values of from 4 to 6. Peaks from vasopressin could be observed by means of an ultraviolet lamp, but were dissolved by the acetic acid destainer (10% acetic acid, 45% ethanol, applied after Coomassie staining). The two-dimensional gels yielded separated precipitates for the BSA and the hormones. Cross-reactions and the possibility of hormones being retained by BSA were not subjected to further examination. Between a pH of 4 and a pH of 7 the optical brightener applied migrated towards the anode.

References

Borner K, Vogt W (ed) (1981) Radioaktivitätsfreie quantitative Immunoassays in der klinischen Chemie. Thieme, Stuttgart, S. 111–122

Clarke HGM, Freeman T (1968) Quantitative immunoelectrophoresis of human serum proteins. Clin Sci 35:403–413

Decker WJ, Nusynowitz ML, Brown CYE, Arnold EC (1968) Purification of antidiuretic hormone by high voltage electrophoresis. J Chromatogr 34:278–281

Laurell CB (1966) Quantitative estimation of proteins by electrophoresis in agarosegel containing antibodies. Anal Biochem 15:45–52

Ohlenschleger G, Berger I, Depner W (1980) Synopsis der Elektrophoresetechniken. GIT Verlag, Darmstadt, S. 124

Rubenstein KE, Schneider RS, Ullman EF (1972) "Homogeneous" enzyme immunoassay. A new immunochemical method. Biochem Biophys Res Commun 74:846–851

Santagostino A, Giagnoni G, Fumagalli P, Pavesi D, Torretta E (1982) Isoelectric point determination of human and camel β-endorphin, α-endorphin, and enkephalins. Biochem Biophys Res Commun 104:577–582

Ullman EF, Schwarzberg M, Rubenstein KE (1976) Fluorescent excitation transfer immunoassay. J Biol Chem 251:4172–4178

Van Weemen BK, Schuurs AHWM (1971) Immunoassay using antigen-enzyme conjugates. FEBS Lett 15:232–236

Vesterberg O (1980) Quantification of proteins with a new sensitive method zone immunoelectrophoresis assay. Hoppe Seylers Z Physiol Chem 361:617–624

Discussion X

Wiegand: Dr. Kamin, did you solve the problem of the reproducible quantification of β-endorphin, or is this a purely qualitative method?

Kamin: Yes, wie need further research to make it a quantitative method. I presented my results just to show that you can create precipitates in this system with an optical brightener. I tried to show the conditions under which this is possible. Electrophoresis is still not sensitive enough.

Wiegand: How can you make an assay of a protein when it is already precipitated?

Kamin: At present I must admit that it is only possible with a standardized substance.

Spintge: If I understood Dr. Kamin correctly, he is looking for a method which is easy to handle and cheaper than the methods we use at the moment. Only very few laboratories in the world measure β-endorphin, because the radio-immunoassay is very expensive. Therefore, I think his starting-point of electrophoresis is very interesting.

Wiegand: Yes, but the problem is that you cannot quantify it.

Spintge: That is correct, but this will be the next step, I think.

Wiegand: I don't see a possibility, because the protein is destroyed. I think with a precipitate you cannot carry out a radio-immunoassay. You cannot make any quantification if the protein is denatured.

Kamin: The precondition for a quantitative measurement should be something similar to the Laurell electrophoresis. The Laurell electrophoresis has precipitates the height of which is proportional to the concentration of the antigen. This would be a quantifiable method. In my study the precipitation takes place in the gel. You can measure the height of the precipitate band. But yet you cannot measure in the range of picograms with the electrophoresis techniques we have today. I tried to show how we can do it by means of glass tubes. This method has been proposed by Westerberg. You measure the distance which the antigen wandered in the antibody-containing gel before it precipitated. This distance is proportional to the concentration of the antigens. This is a ready-made quantitative method, which in some antigens comes down to the nanogram range.

Wiegand: So you would be so optimistic as to assume that with β-endorphin this might also be possible?

Kamin: I see this possibility.

Wiegand: We are very eager to learn your results.

Droh: Dr. Kamin is trying to find a way of setting us free from the complicated measurements with radio-immunoassay.

Kamin: I would like to close with one remark regarding the Westerberg method. With this method there are many variations possible concerning the technical arrangements and the chemical substances. The only thing is that you always have to use agarose. So I think a solution can be found.

β-Endorphin for Treatment of Pain

T. Oyama

Opiate receptors were found in the brains of mammals independently by Pert and Snyder, Simon et al., and Terenius in 1973. These neuronal membrane proteins have been demonstrated to mediate the pharmacological effects of opiate drugs by binding to very specific receptors which in turn produce the biological effects.

β-Endorphin has been isolated from pituitary glands independently by Cox et al. and by Guillemin et al. (1976). β-Endorphin has a 31 amino-acids sequence (61–91) of β-lipotropin, which in turn is a cleavage product of pro-opiocortin, which is referred to as "big ACTH" (265 amino-acids, molecular weight about 31 000), the precursor peptide for ACTH (adrenocorticotropic hormone (Fig. 1).

β-Endorphin, a molecule which occurs naturally in the central nervous system and is most likely involved in the normal physiological mechanisms of pain perception, may be of major clinical interest in managing intractable pain. Previous

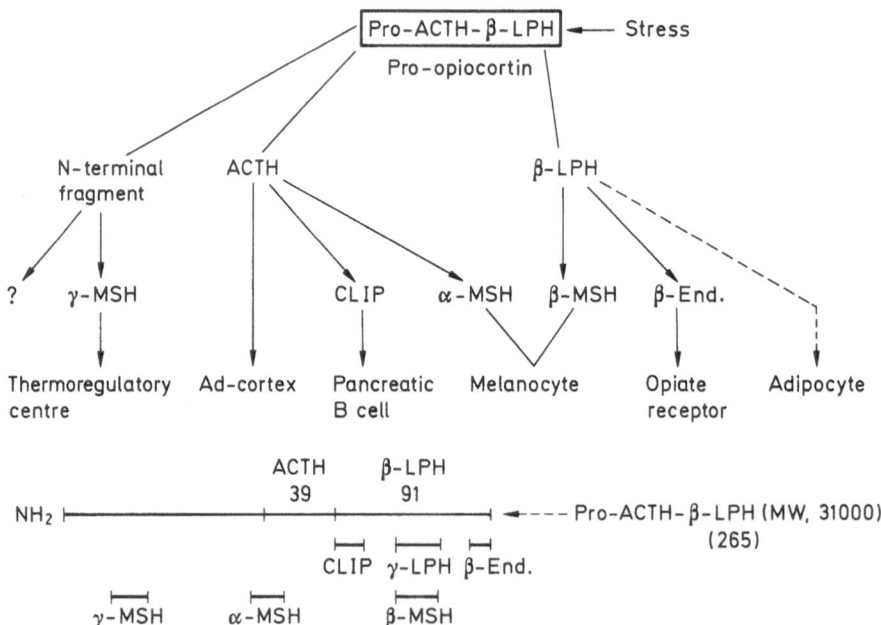

Fig. 1. Production of β-endorphin (*β-End.*). *ACTH*, adrenocorticotropic hormone; *β-LPH*, β-lipotropin; *MSH*, melanocyte-stimulating hormone; *Ad-cortex*, adrenal cortex; *CLIP*, cortico lipotropin; *β-End.*, = β-endorphin

136

reports have demonstrated that the synthetic replicates of the endogenous opioid peptides have profound analgesic effects in laboratory animals when introduced directly into the cerebral ventricles or the brain periaqueductal grey (Feldberg and Smyth 1976). In man intraventricular injection of β-endorphin has also been reported to produce analgesia (Hosobuchi and Li 1978). Hosobuchi et al. (1978) who injected 400 µg β-endorphin into the third ventricle of three cancer patients, observed analgesia which lasted for 8 h. However, no dose of β-endorphin, the most potent of the opioid peptides, has reproducibly induced analgesia when injected intravenously either in laboratory animals or in man (Bloom et al. 1976; Catlin et al. 1977). Catlin et al. (1977) intravenously injected 20 mg β-endorphin in patients, who fell asleep, but there was not always a clearly demonstrated analgesic effect. Foley et al. (1979) could not find any analgesic effect of β-endorphin injected intravenously in patients.

Intrathecal Administration of β-Endorphin for Relief of Cancer Pain

Method

The authors studied the analgesic effect of intrathecally injected β-endorphin in man in August 1979 in collaboration with Prof. R. Guillemin, Salk Institute, San Diego, California. We selected 14 patients with chronic intractable pain in the back, chest, abdomen, and rectal and thigh regions secondary to metastatic malignancies. Systemically administered analgesics had not completely suppressed the pain when given at reasonable dose levels and frequency; the patients usually did not sleep well because of their pain (Oyama et al. 1980a). Synthetic human β-endorphin 3 mg, with 150 mg glucose and 3 ml distilled water or with 3 ml physiological saline and no glucose, was injected intrathecally at the L2-L3 lumbar interspace through Millipore filtration.

Results

All 14 patients reported complete and long-lasting relief from pain. Pain relief was rapid and obtained 1–5 min after the intrathecal injection of β-endorphin. Mean duration of relief of pain after the injection was 33.4 h (range 22.5–73.5 h). The pattern of changes of intensity of pain in these patients is demonstrated in Fig. 2. No effect or a transient effect of the placebo (physiological saline intrathecally) was found in five of the six patients randomly selected for this part of the study (Fig. 2).

The intrathecal injection of β-endorphin caused no discomfort to the patients; 11 of 14 patients became drowsy for 1–4 h after injection; seven patients slept for 1–2 h. All were easily aroused by verbal stimuli. Dysuria was reported by three patients. No signs of respiratory depression, arterial blood gas changes, nausea, hypotension, hypothermia, catatonia, or muscle rigidity were observed. During the period of pain relief, perception of venipuncture and light touch remained intact and voluntary movement was normal. No abnormalities in the ECG or EEG were recorded during the procedure. Minor and transient confusion, mostly in the form of superficial disorientation, was observed in several cases. Several patients became euphoric.

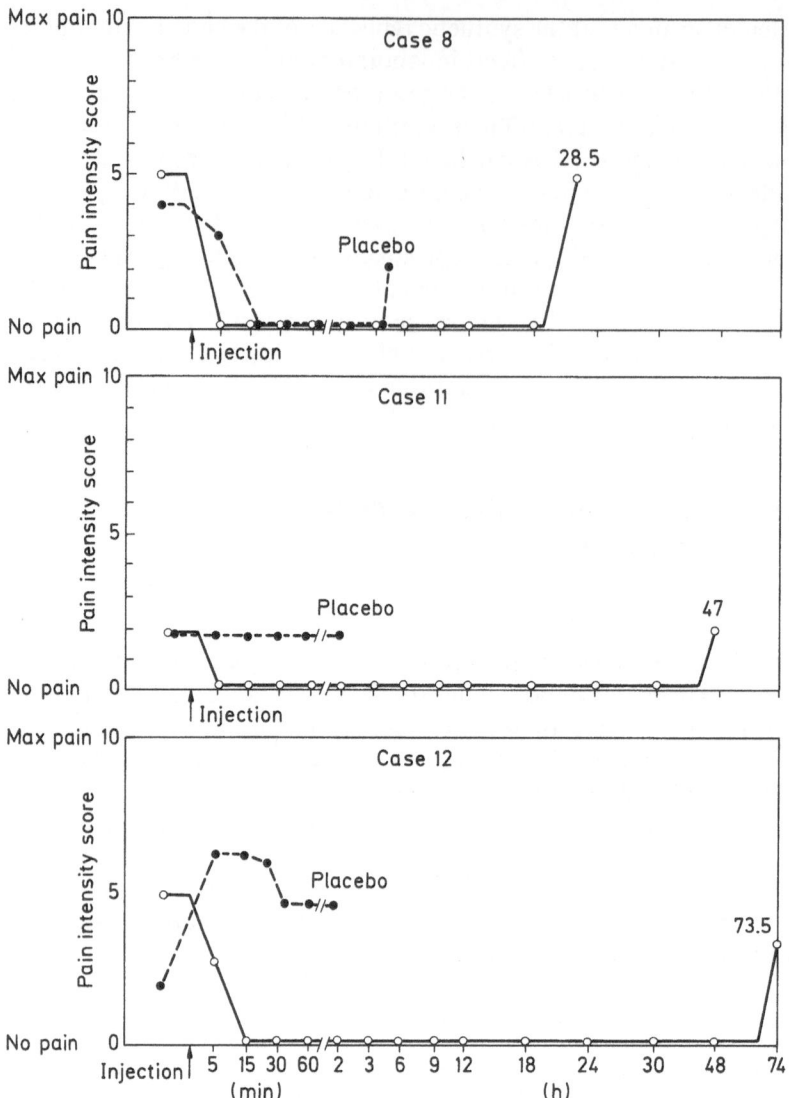

Fig. 2. Pattern of changes of intensity of pain in three of 14 cancer patients

Discussion

Morphine has been injected intrathecally by Wang et al. in eight patients with intractable pain; the onset of pain relief occurred within 15–45 min of administration, and the duration of analgesia (average 15 h, range 12–24 h) was shorter than
that observed here at the doses used (0.5–1.0 mg). Administering morphine at
doses higher than those used by Wang et al. (1979) may not be advisable, since
no difference was seen by these authors between responses to 0.5 and 1.0 mg morphine, and (Wang et al. 1979) higher doses may produce many of the undesirable
effects associated with opiates.

The rapid onset of pain relief after intrathecal injection of β-endorphin suggests
that the opioid peptide acts first at the level of the spinal cord and only later in
the brain after it has diffused throughout the cerebrospinal fluid (CSF). β-Endor-

Table 1. β-Endorphin in obstetric analgesia

Case no.	BW (kg)	Age (y)	Parity	Cervical dilatation (cm)	Onset of analgesia (min)	Injection–delivery time (min)	Neonate BW (g) and sex	Apgar score 1 min, 5 min	Duration of analgesia (h)	Maternal bleeding (g)	Use of oxytocic agent	Maternal BP, PR, ECG, respiration
1	64	23	M	4	2.5	213	3500 M	10, 10	12	100	(+)	Stable
2	57	23	P	4	2.5	88	3200 M	10, 10	15	120	(+)	Stable
3	66	32	M	5	1.9	38	3200 F	10, 10	12	90	(+)	Stable
4	56	27	M	4	2.2	225	3100 F	9, 10	14	150	(+)	Stable
5	55	24	P	8	2.7	24	3460 F	10, 10	15	30	(−)	Stable
6	58	27	P	4	4.0	50	2500 F	9, 10	24	50	(+)	Stable
7	63	25	P	8	1.5	164	3710 M	8, 10	27	100	(−)	Stable
8	62	18	P	10	3.0	87	3300 F	8, 10	15.5	250	(+)	Stable
9	60	31	M	10	4.0	40	3460 M	10, 10	24	500	(−)	Stable
10	64	24	P	9	3.5	23	3170 F	10, 10	32	250	(−)	Stable
11	67	32	M	4	2.3	260	3510 F	8, 10	19	30	(−)	Stable
12	71	30	M	5	6.0	89	3330 M	10, 10	16	100	(+)	Stable
13	49	26	P	8	4.5	19	2400 F	9, 10	22	140	(+)	Stable
14	58	23	M	5	9.0	123	3130 M	10, 10	15	80	(−)	Stable
Mean	60.7	26.1		6.3	3.54	103.0	3212	9.4	18.8	142		
±SE	±1.5	±1.0		±0.61	±0.51	±21.1	±94.5	±0.22	±1.6	±31.8		

BW, body weight; BP, blood pressure; PR, pulse rate; P, primaparous; M, multiparous

phin administered intrathecally also produces cerebral effects, and this is probably the explanation for the drowsiness and the possible transient mild confusion observed in several patients. No EEG abnormalities were observed. It is remarkable that the analgesic response to a single dose of β-endorphin was so consistent in patients in whom very large and frequent doses of opiates had given only partial relief. This suggests that the receptor mechanisms for the two types of substance are not necessarily identical, and this agrees with some current thinking in that field.

β-Endorphin for Obstetric Analgesia

Materials and Methods

Fourteen patients ranging in age from 18 to 32 years in whom normal vaginal delivery was expected were selected as the subjects for our study. They consisted of seven primiparas and seven multiparas with uncomplicated pregnancies who were scheduled for vaginal delivery, either spontaneous or induced (Table 1). Synthetic β-endorphin 1 mg, diluted with 1 ml physiological saline, was injected intrathecally at the third or fourth lumbar interspace with a 22–23 gauge needle at the rate

Case 2 FH 23y 57kg Primipara

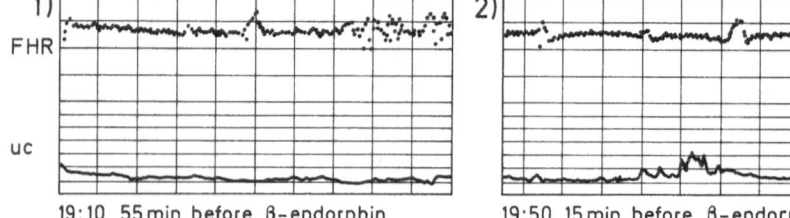

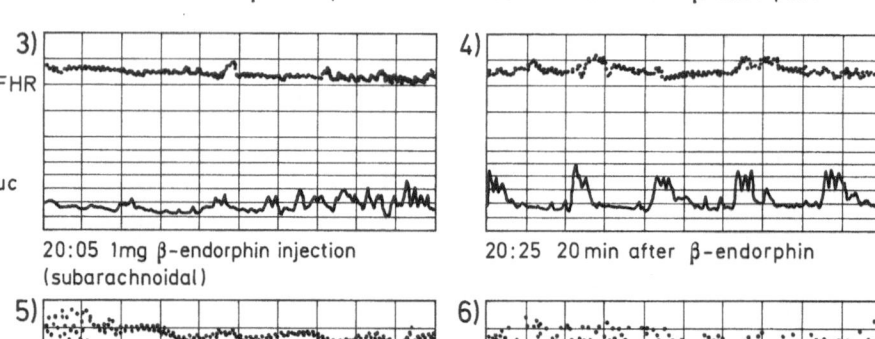

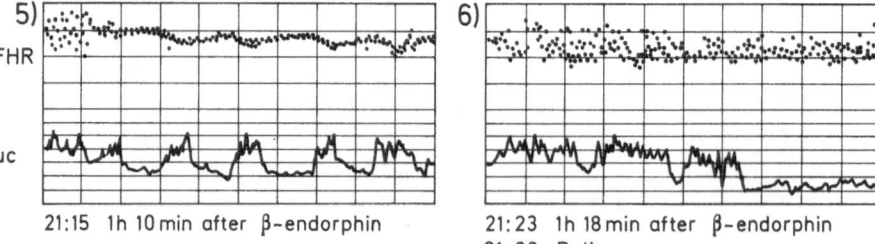

Fig. 3. Foetal heart rate (*FHR*) and uterine contractions (*uc*) before and after administration of β-endorphin for obstetric analgesia (**a**) in a primipara and (**b**) in a multipara

of 0.2 ml/s at the time when cervical dilation was 3.5–5.0 cm, though this was not always possible due to the progress of labour (Oyama et al. 1980 b).

Results

The average cervical dilatation at the time of injection was 6.3 ± 0.6 cm (range 4–10 cm). The time from injection to delivery was 103 ± 21 min, ranging from 19 min to 260 min. Labour pains in all patients disappeared completely within 3.5 ± 0.5 min after the injection of β-endorphin, although feeling of pressure over the lower abdomen and of resistance over the rectovaginal region at the time of the uterine contractions remained unchanged. There was no depression of uterine contractions or slowing of the fetal heart rate due to the administration of β-endorphin (Fig. 3). All Apgar scores were over 8 at 1 min and 10 at 5 min after delivery in all infants (Table 1).

Uterine contraction after delivery was excellent in all patients. All women reported complete absence of perineal discomfort for as 12–32 h after completion of labour (Table 1). Blood gas analyses in the umbilical vein immediately after delivery showed normal levels of PO_2, PCO_2, pH, and base excess.

During the period of pain relief, perception of venipuncture and light touch remained intact, and voluntary movement of extremities was normal with no evi-

Case 3 TH 32y 66 kg Multipara

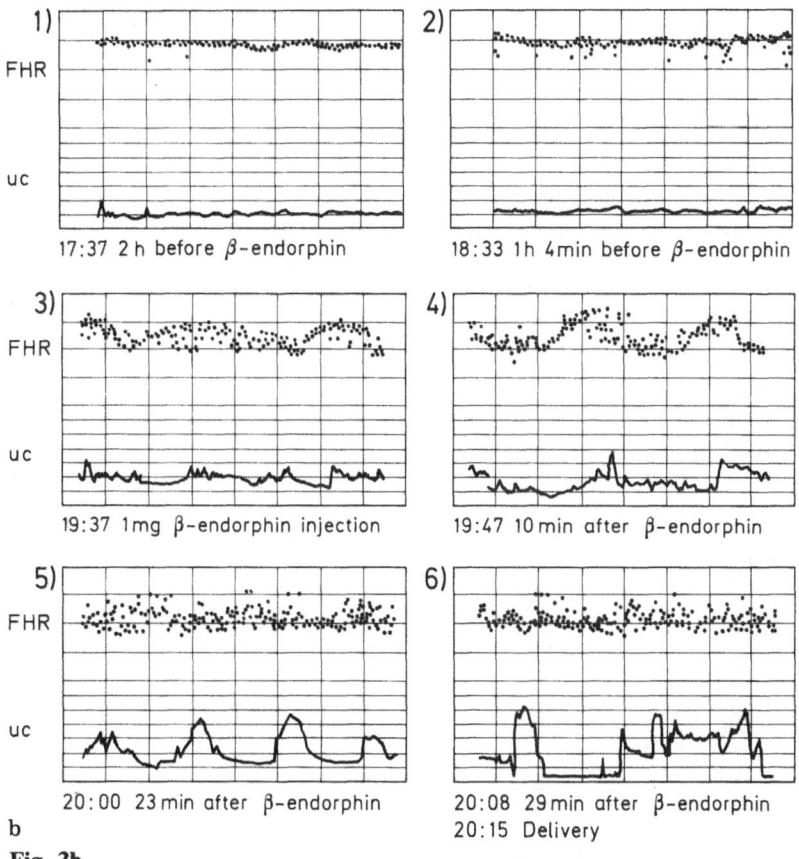

Fig. 3b

141

dence of muscle rigidity or anything resembling catatonia. No hypothermia was observed. All women were alert, although three were slightly drowsy for 30 min to 1 h after the injection of β-endorphin. Nausea and vomiting was noted in four patients and headache in ten; this is most likely related to the size of the lumbar puncture needle and the subsequent possible leakage of CSF; with the use of finer gauge-24 needles in the last four cases no headache was reported.

Discussion

The advantages of β-endorphin as an obstetric and analgesic agent are to be found in the fact that β-endorphin is a normally occurring secretion of the pituitary gland; its degradation products are simple amino-acids. Because it does not pass through the blood-brain barrier, it cannot penetrate to the central nervous system of the foetus, in contradistinction to all other drugs at present used in obstetric analgesia or anaesthesia.

Epidural β-Endorphin

Synthetic β-endorphin 3 mg in 10 ml physiological saline, sterilized by Millipore filter, was injected epidurally at the appropriate lumbar interspace. Ten patients with chronic intractable pain in the back, chest, abdomen, hip, and thigh regions secondary to metastatic malignancies were selected for study.

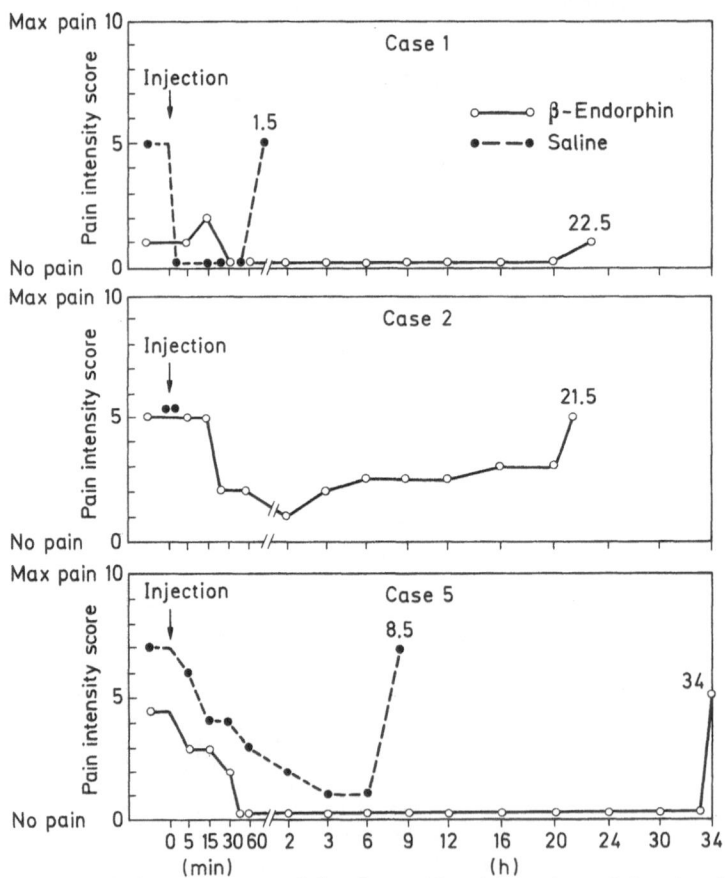

Fig. 4. Pain intensity score following epidural injection of β-endorphin

All ten patients reported relief from pain. Pain relief was obtained within a mean of 24.9 ± 3.4 min (range 9–45 min) after the epidural injection of β-endorphin. This was slower than that produced by the intrathecal injection of β-endorphin 3 mg, which produced rapid pain relief 1–5 min after the injection. Mean duration of relief of pain after the injection was 19.5 ± 3.7 h (range 3.5–35 h), which was shorter than that which had been produced by intrathecal injection (mean 33.4 h; range 22.5–73.5 h) (Oyama et al. 1982).

The pattern of changes of intensity of pain in these patients is demonstrated in Fig. 4. No effect or only a transient one was found after injection of placebo in five out of the seven patients randomly selected for this part of the study. The epidural injection of β-endorphin caused no discomfort, signs of respiratory depression, arterial blood gas changes, hypotension, hypothermia, catatonia or muscle rigidity. Four out of ten patients slept during the procedure.

Complications

Severe complications such as paralysis could occur from intrathecal or epidural injection of narcotics due to preservatives, stabilizers and/or antioxidants used in their preparation. Another problem is that increasing the dosage of intrathecal or epidural narcotic results in only a slightly increased duration of analgesia compared to a lower dose. Besides, higher intrathecal doses have been associated with severe respiratory complications and dysuria.

Conclusion

The intrathecal route is better than the epidural one since:

1. Onset of analgesia is earlier (about 3 min) than with the epidural.
2. Duration of analgesia is slightly longer (30 h) than with the epidural.
3. Relief of pain is complete with the intrathecal route but is only partial with the epidural route.

The comparison between morphine and endorphin is shown in Table 2.

Table 2. Comparison between intrathecally administered morphine and β-Endorphin

	Morphine	β-Endorphin
Analgesia		
Onset	Relatively slow (15–45 min)	Rapid (5 min)
Duration	Shorter (15 h; 12–24 h)	Longer (33.4 h; 22–73 h)
Depression		
Respiratory	+	−
Cardiovascular	+	−
Nausea and vomiting	±	−
Itching	+ +	−
Price	Cheap	Expensive
Tolerance	+ +	?
Withdrawal symptoms	+ +	?

References

Bloom F, Segal D, Ling N, Guillemin R (1976) Endorphins: profound behavioral effects in rats suggest new etiological factors in mental illness. Science 194:630–632

Catlin DH, Hui KK, Loh HH, Li CH (1977) Pharmacologic activity of β-endorphin in man. Commun Psychopharmacol 1:439–500

Cox BM, Goldstein A, Li CH (1976) Opioid activity of a peptide, β-lipotropin-(61–91), derived from β-lipotropin. Proc Natl Acad Sci USA 73:1821–1823

Feldberg W, Smyth DG (1976) The C-fragment of lipotropin – a potent analgesic. J Physiol (Lond) 260:30–31

Foley KM, Kourides IA, Inturrisi CE et al. (1979) β-Endorphin: analgesic and hormonal effects in humans. Proc Natl Acad Sci USA 76:5377

Hosobuchi Y, Li CH (1978) The analgesic activity of human β-endorphin in man. Commun Psychopharmacol 2:33–37

Lazarŭs LH, Ling N, Guillemin R (1976) β-Lipotropin as a prohormone forth morphinomimetic peptides endorphins and enkephalins. Proc Natl Acad Sci USA 73:2156–2159

Oyama T, Fukushi S, Jin T (1982) Epidural β-endorphin in treatment of pain. Can Anaesth Soc J 29:24–26

Oyama T, Jin T, Yamaya R, Ling N, Guillemin R (1980a) Profound analgesic effects of β-endorphin in man. Lancet I:122–124

Oyama T, Matsuki A, Taneichi T, Ling N, Guillemin R (1980b) β-Endorphin in obstetric analgesia. Am J Obstet Gynecol 137:613–616

Pert CB, Kuhar MJ, Synder SH (1976) Opiate receptors: autoradiographic localization in rat brain. Proc Natl Acad Sci USA 73:3729–3733

Simon EJ, Hiller JM, Edelman I (1973) Stereospecific binding of the potent narcotic analgesic [^{3}H]etorphan to rat-brain homogenate. Proc Natl Acad Sci USA 70:1947–1949

Terenius L (1973) Characteristics of the "receptor" for narcotic analgesics in synaptic plasma membrane fraction from rat brain. Acta Pharmacol Toxicol 33:377–384

Wang JK, Naus LA, Thomas JE (1979) Pain relief by intrathecally applied morphine in man. Anesthesiology 50:149

Discussion XI

Wiegand: Prof. Oyama, how did you measure β-endorphin?

Oyama: With a radio-immunoassay.

Wiegand: Did you see any interference with β-lipotropin?

Oyama: This very special problem should be discussed afterwards.

Involvement of Pituitary Endorphins in Pain Perception and Their Importance in Pituitary-Stimulation-Induced Analgesia in Animals and Man

A. Trouwborst, H. Yanagida, and W. Erdmann

Introduction

Recently, we suggested that the pituitary gland possibly plays a major role in pain perception. Experimental findings by various investigators have supported this idea:

1. In animals severe stress induces analgesia concomittant with a simultaneous increase in the hormonal activity of the pituitary gland (Guillemin et al. 1977; Lewis et al. 1980; Willer and Albe-Fessard 1980; Rossier et al. 1977).
2. After surgical hypophysectomy stress-induced analgesia is absent (Vidal et al. 1982; Amir and Amit 1979; Bodnar et al. 1979).
3. After injection of the opiate (endorphin) antagonist naloxone, stress-induced analgesia is absent (Chesher and Chan 1977; Bodnar et al. 1978).
4. Other investigators have found high levels of endorphins in the neurohypophysis, with endorphin receptors present and even enkephalinergic neurons in the pituitary region (PR) (Goldstein 1976; Bloom et al. 1978; Simantov and Snijder 1977; Rossier et al. 1980).

Further support for a major general role of the PR in pain perception was added by Moricca: injection of alcohol into the PR caused relief of intractable cancer pain not only in cases with hormone-dependent tumours but also in cases with hormone-independent tumours (Moricca 1974). This treatment involving just the PR has meanwhile been applied in more than 10000 patients with great success.

The primary explanation that the pain-relieving effect of the so-called neuroadenolysis of the pituitary gland was due to gland destruction was found to be wrong. Histological studies of the PR in monkeys after alcohol injection which was followed by pain relief showed only partial destruction of the pituitary tissue (Trouwborst 1982). The degree of tissue destruction had no influence on the reduction of the evoked potential in the primary sensory cortex (PSC), and even with minor tissue damage full depression of the PSC evoked potential occurred. In recent experimental studies simultaneous recording of neuronal activity in the PR was introduced. The "neurohypophysis" showed a brain-cortex-like neurophysiological behaviour, and characteristic evoked potentials could be obtained in response to tooth pulp stimulation simultaneously with the routinely registered evoked potentials in the PSC. Injection of alcohol into the PR was followed by an increase in the tooth pulp evoked potential (TPEP), whilst the already known decrease in the TPEP was observed in the PSC. Administration of naloxone, an endorphin antagonist, after alcohol neuroadenolysis reversed the effects; TPEP in the PR and TPEP in the PSC returned towards normal with reappearance of pain sensation as expected (Trouwborst 1982).

The observations described above ultimately led to the following assumptions:

1. Pain relief after alcohol injection into the PR was due to "activation of the PR through wounding".
2. An increase in the neurohypophyseal activity had an inhibitory effect on the PSC activity and its behaviour in response to a pain stimulus.
3. The neurohypophysis-stimulating effect of alcohol injection was partially due to increased endorphin activity, as it could be blocked to a great extent by endorphin antagonists.

The purpose of the present study was to get further insight into the role of the PR in pain perception by investigation of:

1. The effect of naloxone in intact animals on the pituitary neuronal activity compared to the effects of naloxone on the PSC activity.
2. The effects of an opiate in imitating the assumed endorphin-mediated hyperactivity of the pituitary gland.
3. Direct electrical stimulation of the neurohypophysis to imitate the assumed endorphin-mediated hyperactivity of the pituitary gland.

Methods

Tooth pulp stimulation was used to produce acute experimental pain (Chabrian et al. 1975; Tachibana 1975; Chen et al. 1974; Guig et al. 1981). The neurophysiological responses (evoked potentials) in the PSC were registered. The attitude of the TPEPs recorded from the PSC is known to be proportional to the stimulus intensity and is correlated to subjective pain sensation in a quantitative manner. The simultaneous recording of evoked potentials in the PR was recorded as well; this has, according to our information, never been done before.

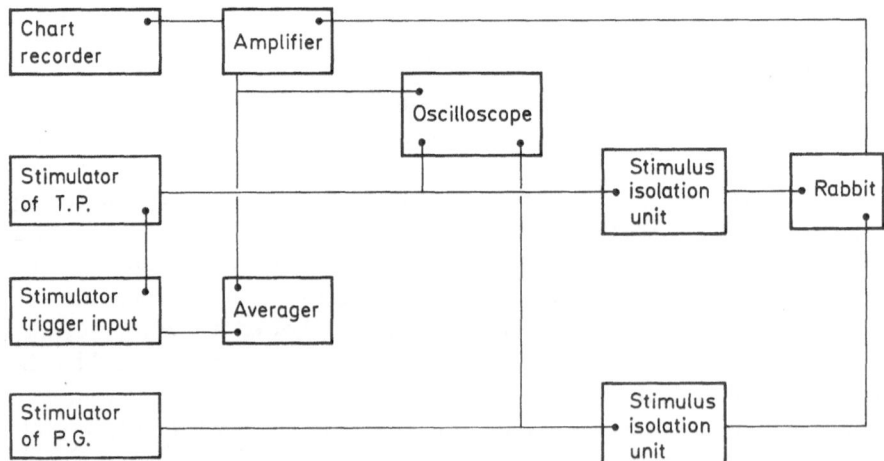

Fig. 1. The experimental set-up. *TP*, tooth pulp; *PG*, pituitary gland

Under anaesthesia the stimulating electrodes were introduced into the tooth pulp of rabbits and monkeys via small bore holes and fixed by dental acrylic. Recording electrodes were placed via bore holes on to the PSC and into the PR via the nasopharyngeal route. In fully awake animals tooth pulp stimulation was applied and the response (TPEP) was recorded in the PSC and the PR. By means of an averager 50 single TPEPs were averaged under each condition and the averaged TPEP was printed out on an X-Y plotter (Fig. 1).

In a first series of six rabbits naloxone, an opiate antagonist, was investigated for its effects on the TPEP in both areas. In a second group of six rabbits fentanyl, an opiate agonist, was administered. In a third series of experiments electrical stimulation was applied to the electrodes in the PR while TPEPs were recorded from the PSC. At a later stage all experiments were repeated in three monkeys to verify that the results found in rabbits could be reproduced in the primate.

Results

After intravenous injection of naloxone the TPEP in the PSC was increased (hyperalgesia). At the same time the response in the PR was decreased, probably as a result of naloxone inhibition of endorphin-mediated transmission. Thus hyperalgesic activity of the PSC is most probably produced via endorphin antagonism in the PR (Fig. 2).

Intravenous injection of fentanyl induced a decrease in TPEP in the PSC (hypoalgesia), as expected, and an increase in the response in the PR. Thus a possible means of action of opiates is through activation of the PR, which leads to an inhibition of pain sensation (Fig. 3).

After a repeated electrical stimulation of the PR for 10 min, the TPEPs in the PSC were severely depressed for several hours (Fig. 4). After electrical stimulation followed by injection of naloxone, the electrical stimulation effect was partially reversed, with a recovery of the TPEPs in the PSC by ±25%.

The control experiments in monkeys showed the same results as those in rabbbits.

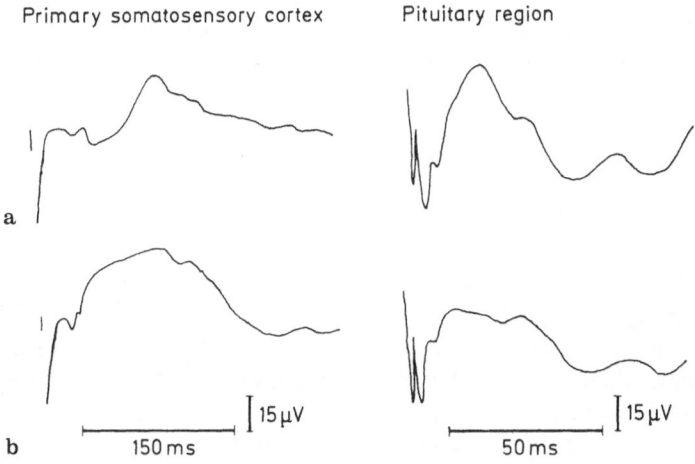

Primary somatosensory cortex   Pituitary region

15 µV   150 ms   15 µV   50 ms

Fig. 2a, b. The influence of naloxone on the TPEP of the PSC and of the PR. **a** before naloxone; **b** after naloxone

Primary somatosensory cortex   Pituitary region

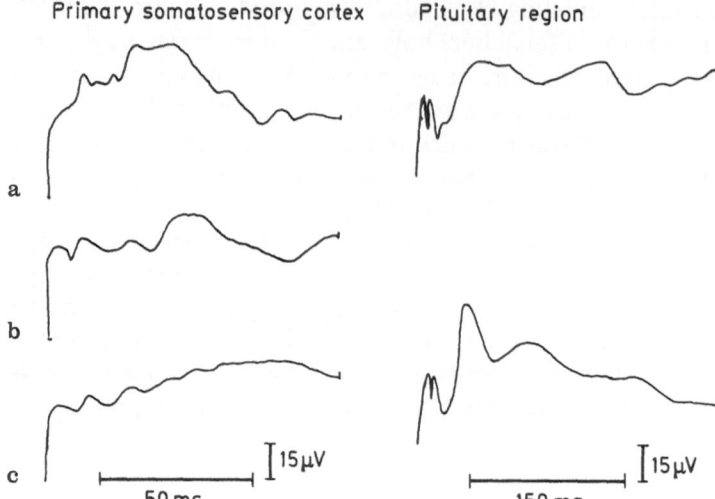

Fig. 3a-c. The influence of fentanyl on TPEP of the PSC and of the PR. **a** before fentanyl; **b** 5 min after fentanyl; **c** 10 min after fentanyl

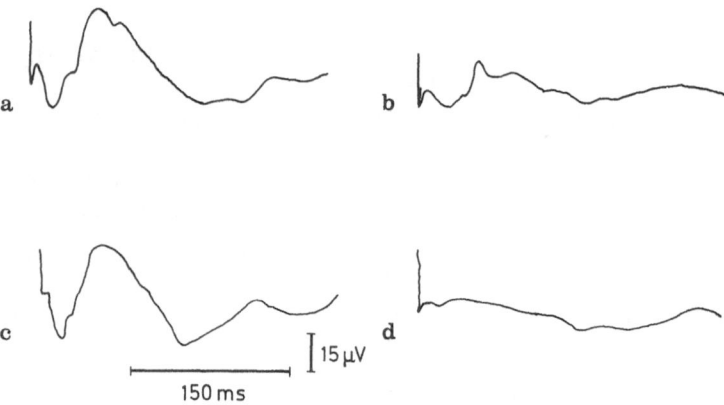

Fig. 4a-d. The effect of stimulation of the PR on the TPEP of the PSC. **a** before stimulation; **b** after stimulation; **c** 1 h after stimulation; **d** after the second stimulation period

Conclusions

Activation of the PR (by local alcohol injection, intravenous opiates or electrical stimulation) is followed by pain relief.

The results of the fentanyl and naloxone studies suggest that pituitary endorphin receptors and enkephalinergic pathways are partially involved. The PR seems to be a switchboard in the pain perception pathways, whereby pituitary endorphins seem to play a major role. To summarize: activation of the PSC occurs through inhibition of the PR, inhibition of pain perception (PSC) through activation of the PR.

For treatment of intractable cancer pain, opiates and/or alcohol adenolysis of the pituitary gland can be replaced by electrical stimulation of the PR. This avoids the specific side-effects of high-dosage opiate administration (e.g. respiratory depression, deep sedation) or the dangers of intrapituitary alcohol injections (damage of surrounding vital structures).

Meanwhile, electrical stimulation has been used for treatment of intractable cancer pain (Yanagida et al., unpublished) in a repidly increasing number of Japanese patients with excellent results. So far, no complications have occurred.

References

Amir S, Amit Z (1979) The pituitary gland mediates acute and chronic pain responsiveness in stressed and non-stressed rats. Brain Res 190:102–107

Bloom FE, Rossier J, Battenburg ELF et al. (1978) β-Endorphin: cellular localization, electrophysiological, and behavioural effects. Adv Biochem Psychopharmacol 18:89–107

Bodnar RJ, Kelly DD, Spiaggia A et al. (1978) Dose dependent reduction by naloxone of analgesia induced by cold-water stress. Pharm Biochem Behav 8:667–672

Bodnar RJ, Glusman M, Bruters M et al. (1979) Analgesia induced by cold-water stress: attenuation following hypophysectomy. Physiol Behav 23:52–62

Chabrian GE, Canfield RC, Knauss FA, Lettich E (1975) Cerebral responses to electrical tooth pulp stimulation in man. Neurology 25:745–757

Chen ACN, Chapman CR, Harkins SW (1974) Brain evoked potentials are functional correlates of induced pain in man. Pain 6:365–374

Chesher GB, Chan B (1977) Footshock induced analgesia in mice: its reversal by naloxone and cross tolerance with morphine. Life Sci 21:1569–1574

Goldstein A (1976) Opioid peptides (endorphins) in pituitary and brain. Science 193:1081–1086

Guig JD, Colpits YH, Chapmann CR (1981) Effects of local anesthetic infiltration on brain potentials evoked by painful dental stimulation. Anaesth Analg 60:779–783

Guillemin R, Vargo T, Rossier Y (1977) β-Endorphin and adrenocorticotropin are secreted concomitantly by the pituitary gland. Science 197:1367–1369

Lewis JW, Cannon JT, Liebeskind JC (1980) Opioid and nonopioid mechanisms of stress analgesia. Science 208:623–625

Moricca G (1974) Chemical hypophysectomy for cancer pain. Adv Neurol 4:707–714

Rossier J, French ED, Rivier C et al. (1977) Foot-shock induced stress increases β-endorphin levels in blood but not in brain. Nature 270:618–620

Rossier Y, Pittman Q, Guillemin R (1980) Distribution of opioid peptides in the pituitary: a new hypothalamic – pars nervosa enkephalinergic pathway. Fed Proc 39:2255–2260

Simantov R, Snijder S (1977) Opiate receptor binding in the pituitary gland. Brain Res 124:178–184

Tachibana N (1975) Somatosensory evoked potentials and analgesia in man. Int Anaesth Clin 13(1):191–201

Trouwborst A (1982) The role of the pituitary region in the endogenous pain control mechanism. PhD Thesis, Erasmus University, Rotterdam

Vidal C, Girault JM, Jacob J (1982) The effect of pituitary removal on pain regulation in the rat. Brain Res 223:53–64

Willer JC, Albe-Fessard D (1980) Electrophysiological evidence for a release of endogenous opiates in stress-induced analgesia in man. Brain Res 198:419–426

Sufentanil: A Synthetic Narcotic
for Total Intravenous Anaesthesia?

D. M. Philbin[1], *C. E. Rosow, R. C. Schneider,*
M. D'Ambra, E. Freis, and V. Machaj

Introduction

High-dose fentanyl anaesthesia has been reported to suppress or at least attenuate
the increases in stress hormones associated with surgical stimulation (Sebel et al.
1981; Stanley et al. 1980). Sufentanil, a potent analogue of fentanyl, has more affinity for opiate receptors (Stahl et al. 1977) and a higher therapeutic index (De
Castro et al. 1979), and thus may be more efficacious in blocking these hormonal
responses. A recent study (Bovill et al. 1983) demonstrated that sufentanil in a
dose of 15 µg/kg could not block the hormonal increases which occurred during
cardiopulmonary bypass. This study was undertaken to compare the hormonal
response to these drugs in the clinical setting.

Methods

Twenty patients scheduled for elective coronary artery surgery were selected for
study. Informed consent was obtained in each case and the patients were randomly divided into four groups according to narcotic dose: group 1, 50 µg fentanyl per kilogram; group 2, sufentanil 10 µg/kg; group 3, fentanyl 100 µg/kg;
group 4, sufentanil 20 µg/kg. All narcotics were administered following a double-
blind protocol. Routine monitoring devices were positioned under local anaesthesia prior to induction. These included central venous and pulmonary artery
thermodilution catheters, as well as an indwelling radial artery catheter and leads
II and V_5 of the electrocardiogram. Monitoring throughout the study was continuous via an eight-channel recorder with an oscilloscope.

Haemodynamic measurements and blood samples for aldosterone, cortisol and
catecholamine levels were obtained prior to the start of anaesthesia, after intubation, after skin incision, after sternotomy, during hypothermic bypass, 1 h past
bypass, and in the surgical intensive care unit at the end of the procedure.

Results

The patients in the four groups were comparable for age, sex, type of operation,
length of operation and the degree of pre-operative cardiac dysfunction. Haemodynamic data for the four groups of patients followed a remarkably similar pattern. Cardiac output gradually declined in all groups. The mean and systolic

[1] Supported in part by Senior International Fellowship No. 1F06 TW00705-01
Fagarty Center, National Institutes of Health

Table 1. Cortisol and aldosterone concentrations expressed as percentage of control

Period	Group 1		Group 2		Group 3		Group 4	
	Aldo (pg/ml)	Cortisol (pg/ml)	Aldo (pg/ml)	Cortisol (pg/ml)	Aldo (pg/ml)	Cortisol (pg/ml)	Aldo (pg/ml)	Cortisol (pg/ml)
Intubation	79	63	87	98	110	72	100	91
Incision	113	49	99	84	67	58	113	86
Sternotomy	128	164	82	60	77	76	105	98
30 min on CPB	103	73	122	125	106	66	105	105
Rewarming on CPB	163	105	154	145	113	85	160	92
1 h after CPB	152	102	240	157	158	115	195	96
SICU	151	101	260	201	154	111	246	193

CPB, cardiopulmonary bypass; SICU, surgical intensive care unit

Table 2. Epinephrine and norepinephrine concentrations expressed as percentage of control

Period	Group 1		Group 2		Group 3		Group 4	
	Epi (pg/ml)	Norepi (pg/ml)	Epi (pg/ml)	Norepi (pg/ml)	Epi (pg/ml)	Norepi (pg/ml)	Epi (pg/ml)	Norepi (pg/ml)
Intubation	272	115	76	75	62	94	43	81
Incision	528	118	71	51	87	85	38	84
Sternotomy	193	101	114	95	98	103	71	90
30 min on CPB	883	203	852	211	233	173	177	92
Rewarming on CPB	1048	333	322	175	169	176	194	233
1 h after CPB	493	213	212	105	183	147	346	334
SICU	582	258	242	144	151	93	252	394

CPB, cardiopulmonary bypass; SICU, surgical intensive care unit

blood pressures demonstrated slight increases following sternotomy in groups 1 and 2 (126–133 mm Hg and 120–136 mm Hg) but not in groups 3 and 4. However, there were no statistically significant diferences between the groups.

Table 1 contains the cortisol and aldosterone concentrations throughout the period of the study expressed as a percentage of control. The initial decrease in cortisol occurred in all four groups and was followed by increases which achieved significance only in the patients receiving sufentanil (groups 2 and 4). Aldosterone concentrations followed a similar pattern but the increases did not achieve significance in group 3 (fentanyl 100 µg/kg).

Table 2 contains the epinephrine and norepinephrine responses for all four groups, again expressed as a percentage of control. Significant increases in epinephrine occurred consistently in group 1 (fentanyl 50 µg/kg). The increases in all groups were significant during the period of cardiopulmonary bypass.

Discussion

It is apparent from these data that the haemodynamic profile alone, in the clinical setting, shows no clear advantage for either drug or dosage level. The similarity of response in the four groups is apparent. This may be in part due to the small numbers of patients involved. If one looks at the pattern of hormonal response, however, a different picture emerges. The clear and significant increases in both cortisol and aldosterone which occurred in group 2 (sufentanil 10 µg/kg) suggest it was least effective in blunting these hormonal responses. However, for the catecholamine response group 1 (fentanyl 50 µg/kg) was least effective. Groups 4 and 1 have a somewhat similar pattern, and on the basis of the hormonal responses group 3 (fentanyl 100 µg/kg) clearly produced the most attenuation of all hormones except during the period of bypass. These data suggest then two important findings:

1. The haemodynamic response of the patient to surgical stimulation does not reflect the hormonal stress response.
2. Even the highest dose of sufentanil (20 µg/kg) could not eliminate the hormonal responses to bypass.

The evidence would suggest, then, that increasing doses of ever more potent narcotics do not produce complete suppression of hormonal responses, nor do they produce dose-related suppression of cardiovascular reflexes.

References

Bovill JG, Sebel PS, Fiolet JWT, Touber JL, Kok K, Philbin DM (1983) The influence of sufentanil on endocrine and metabolic responses to cardiac surgery. Anesth Analg 62:391–397

De Castro J, Van De Water A, Wouters L, et al. (1979) Comparative study of the cardiovascular, neurological and metabolic side effects of eight narcotics in dogs. Acta Anesthesiol Belg 30:5–99

Sebel PS, Bovill JG, Schellekens APM, Hawker CD (1981) Hormonal responses to high-dose fentanyl anesthesia. Br J Anaesth 53:941–948

Stahl KD, Van Bever W, Janssen P, Simon EJ (1977) Receptor affinity and pharmacologic potency of a series of narcotic analgesic, antidiarrheal and neuroleptic drugs. Eur J Pharmacol 46:199–205

Stanley TH, Berman L, Green O, Robertson D (1980) Plasma catecholamine and cortisol responses to fentanyl-oxygen anesthesia for coronary artery operations. Anesthesiology 53:250–253

Discussion XII

Siedlecky: Prof. Philbin, do you have any information about the concentration of analgesics in the plasma during your measurements? It is interesting to see that you were able to block the stimulus of intubation but not the relatively small surgical stimulus during cardiopulmonary bypass. Perhaps this could be due to the pharmacokinetics of fentanyl, because when you give a high dose of fentanyl the effect becomes minimal after 40 min. In your study the bypass was made 60 min after your injection of sufentanil.

Could you please comment also on the possible effect of dilution in the heart-lung machine?

Philbin: There are some studies on the pharmacokinetics of sufentanil where the plasma level is surprisingly similar to that of fentanyl. I disagree with one point you made. When a large dose of fentanyl is given at the beginning you can have a sufficient anaesthesia for 4–6 h. The blood level falls within 45 min, but it does not go below the analgesic limit. And then it is a very slow tail-off for about 4–6 h. Sufentanil behaves on a very similar pattern.

We also studied the effects of the dilution by the priming solution of the heart-lung machine during the bypass. Generally, it is an insignificant factor. There is an initial drop, but within minutes values are back to the initial levels.

I don't think the stimulation which we see with the bypass is a surgical stimulation. I think it is due to the totally unphysiological status of the patient with artificial circulation and cooling.

Oyama: On the contrary, I think that cardiopulmonary bypass is a very strong stimulus. We studied plasma concentration of β-endorphin and β-lipotropin (β-LPH) during bypass surgery. It was surprising to see how high the plasma levels were. So my hypothesis is now that even if we use a high dose of narcotic it would be very difficult to block the hormonal reaction to stressful stimulation. That is my experience. So we will have to find out the optimal hormonal level and the optimal narcotic level to block the reaction of surgical stimulation without creating new side effects.

Philbin: I agree. Another surprising thing which we found was that the hormonal reaction did not follow the cardiovascular reaction. We had tremendous rises in catecholamine levels, but the patient was not hypertensive. This makes me ask again: Why are we trying so hard to suppress the catecholamine secretion? Perhaps there is a good reason for it. Maybe we should not try to block this reaction!

On a New Topical Anaesthesia for Intra-uterine Manipulations

M. R. van Santen

Introduction

Gynaecological practice often involves intra-uterine manipulations such as intra-uterine device (IUD) insertion or removal and endometrial biopsy for histological diagnosis of disorders causing bleeding. However, especially in the nulliparous woman these procedures are sometimes rather painful. A syncope may even be provoked and cardiac arrhythmias have been recorded during such procedures (Aznar et al. 1976).

In patients with a history of fainting easily, in nervous or nulliparous women, or if a previous attempt at IUD insertion has failed, gynaecologists sometimes apply a paracervical block (PCB) for local anaesthesia (Fig. 1). The PCB technique is linked with potentially hazardous effects when the anaesthetic is absorbed rapidly or is accidentally injected intravenously, resulting in systemic effects such as cardiac arrhythmia or worse. A lethal outcome of such procedures has been reported (Berger et al. 1974).

A new topical intra-uterine anaesthetic technique has been proposed, employing a tiny transcervical catheter (Fig. 2) fixed on a 5-ml tuberculin syringe to flush the endometrial cavity prior to intra-uterine procedures and so anaesthetize the

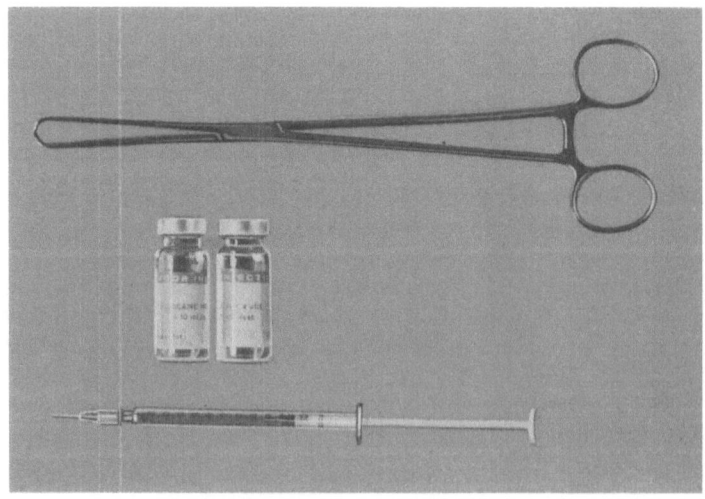

Fig. 1

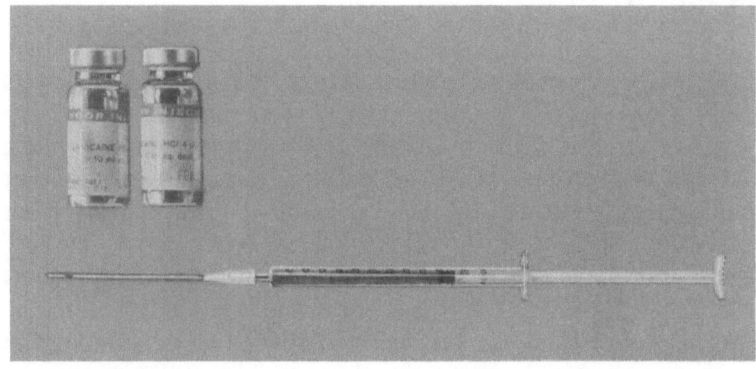

Fig. 2

uterine lining. Painless IUD insertion was reported in 75% of nulliparous women after instillation of a 1%–2% solution of lidocaine 1–2 min before the procedure took place (Hasson 1977). Later, more elaborated studies showed that when the instrument was left in place for 5 min to maintain the injected anaesthetic solution in contact with the uterine mucous membrane, 82% of 443 patients felt no pain, and a further 11% experienced some analgesic effect. In these series 55% of the women were multiparous. In nulliparous women mild and moderate pain and discomfort were noted by 19.7% and 4.4% respectively (Hasson 1982).

Such a technique seemed to be very useful in non-interval IUD insertion, and postcoital IUD insertion in particular, since this is mostly requested by nulliparous women, and in a phase of the menstrual cycle in which the cervical ostium is narrow and rigid (Van Santen and Haspels 1981). Since in the nulliparous the volume of the uterine cavity is limited to 1 ml, we concluded that when 5 ml is used and the cervix is properly sealed off, the tubes are rinsed and part of the solution must be deposited intraperitoneally. There is thus a potential risk of inducing infection in to the internal genital tract. Though in a fertility workup this is a routine procedure carried out under strictly aseptic conditions, such a technique is not suitable for routine use on a large scale. Moreover, nulliparous women tend to be more prone to infections accompanying IUD use than multiparous (Weststrom et al. 1976), and since IUD insertion leads to a 100% contamination of the uterine cavity (Wolf and Krieger 1980), a modified study was designed avoiding such tubal flushing.

Design of the Study

In our modification of Hasson's technique, the amount of flush solution was reduced to 1 ml and an endocervical catheter was employed, as used for intra-uterine insemination. The use of a tuberculin syringe ensures that a limited volume is used and allows easy administration, since it is longer and slimmer than a usual 1-ml syringe. Before any attempt to introduce the catheter into the uterus is made, the fixation site for the vulsellum is prepared by intradermal injection of 0.1 ml of the same anaesthetic fluid at 1 and 11 o'clock in the cervix.

Material and Methods

It is debated among senior gynaecologists whether there is any need for such an anaesthetic. Complications on IUD insertion are said to be rare and never "serious". A slight vagal reaction is easily overcome or better prevented by gentle manipulation, and optional atropine can be added by intramuscular route, they say. However, nulliparity is regarded by them a relative contra-indication for IUD insertion.

This being the case, there was an ideal opportunity in our department to start a double-blind, randomized study on the efficacy of this new modified technique; in a control group the uterine cavity was rinsed with normal saline. Similarly in the control group the intradermal injections for the vulsellum at the cervix were with normal saline.

After routine procedures for preparation of IUD insertion, all our nulliparous patients were offered the opportunity of participating in this study. Fully informed verbal consent was obtained from 53 nulliparous women, and these are represented in this report. In all women an aqueous solution of 4% lidocaine or normal saline was used, distributed in a randomized, double-blind manner. After bimanual gynaecological examination, a vaginal bivalve retractor was inserted in to the vagina, and the cervix centred. The forceps were applied at the cervix as described above, and the cervical catheter was then inserted. During and after the flushing of the uterine cavity it was ensured that the acorn was close against the cervix by applying slight pressure to seal the uterus and prevent any loss of solution to the vagina. The device was left in situ for 1 min before being withdrawn. After assessment of the uterine length, the IUD was inserted as in the routine procedure.

Data Collection

Parameters such as blood pressure and heart rate were measured and subjective data with respect to discomfort, pain and feelings of faintness were recorded. All parameters were measured four times: in the consulting-room after the patient had been informed about the study and the procedure to be undertaken; with the patient in the supine position on the gynaecological examination table prior to the procedure; with the patient in the same position 1 min after the insertion was completed; and in the consulting-room approximately 5 min afterwards.

Compared to the basic results obtained before the procedure, alterations in systolic blood pressure by 15 mm Hg or more at 1 and 5 min after the procedure were observed. A decrease or increase in heart rate above 10 beats/min was registered in the same way. Subjective complaints of pain were counted as positive if the pain was claimed to be more then minimal.

Results

As is shown in Table 1, the systolic blood pressure was lower in the experimental group, but this difference was not statistically significant. The heart rate, however, was reduced more frequently, and compared to the control group this was

Table 1. Results in control and experimental groups

	Normal saline (placebo) ($n=28$)	Lidocaine 4% aq. sol. ($n=25$)	Fishers exact test (two-sided p value)
Systolic blood pressure reduced by ≥ 15 mmHg	8	16	NS ($p=0.09$)
Heart rate reduced by ≥ 10 beats/min	1	5	$p<0.01$
Complaints at			
1 min	19	11	NS
5 min	12	11	NS

statistically significant. The differences in subjective effects such as pain and cramps were small, and not indicative for evaluation.

Discussion

In the initial preliminary report the deposited lidocaine solution was allowed 1 or 2 min to anaesthetize the neural ends in the uterus (Hasson 1977). In the latter study 5 min was the time waited before the procedure was undertaken. Not only was an analgesic effect on the uterine lining expected, but in combination with the deposited lidocaine if 5 ml solution was used, through the flushing out of the tubes and consequently resorption out of Douglas' pouch, a direct anaesthetic effect on the pudendal nerves and uterine branches in the lower pelvis was very likely. In this study, which employed only 1 ml and restricted it to the uterine cavity, such an effect was not possible, and might explain the different outcome.

Though as yet only a limited number of series has been presented, a significant effect on the heart rate in the experimental group has been demonstrated. Rapid absorption by the uterine lining and systemic venous transport induce beta-blockade of the cardionector, followed by bradycardia. Proper application is therefore documented in this new technique. However, the lag-time of the neural ends might be longer than the 1-min waiting time allowed between flushing and IUD insertion in this study. So far no proper anaesthetic effect has been demonstrated. Currently, therefore, a new study is being performed with the same protocol but allowing a full 5 min time to investigate whether this is the clue to this unsatisfactory result.

Another possibility is that the anatomical "bottle-neck", the internal cervical ostium uteri, is not properly anaesthetized by the returning flow of the lidocaine 4% aqueous solution from the uterine cavity to the cervical canal. This trigger point for pain and vagal reactions is apparently not blocked, and this is also possibly due to the fact that there is no deposition of anaesthetic solution in Douglas' pouch.

Another attempt to find an alternative to the paracervical block in IUD insertion is represented by an injection technique which has recently been evaluated (Hepburn 1980; McKenzie and Shaffer 1978). So far more intense pain and more severe bleeding have been observed with this method, though overall only a few patients suffered any pain (Kurz and Meier-Oehlke 1983).

Acknowledgements. The co-operation of the physicians of the Dutch Clinics for Family Planning and Sexuality (Dr. J. Rutgers Stichting), H. Teeuw, H. Doppenberg, B. Best, and of Dr. D. Sojo, gynaecologist from the Madrid Family Planning Service *(Centro Municipal de Planificación Familiar)* in Spain, who participated with the author in this study, is appreciated. The assistance of Mr. G. Eilers and Mr. C. Schmitz of the Department of Biostatistics, Erasmus University, Rotterdam is kindly acknowledged.

References

Aznar R, Reynose L, Ley E, Gámez R, De Leon MD (1976) Electrocardiographic changes induced by insertion of an intrauterine device and other manipulations. Fertil Steril 27:92–96

Berger GS, Tyler CW, Harrod EK (1974) Maternal death associated with paracervical block anaesthesia. Am J Obstet Gynecol 118:1142–1143

Hasson HM (1977) Topical uterine anaesthesia, a preliminary report. Int J Gynaecol Obstet 15:238–240

Hasson HM (1982) Topical uterine anaesthesia for IUD insertion. Contracept Deliv Syst 3:3–4 (Abstract 192)

Hepburn S (1980) Method of local anaesthesia for IUD insertion. Contraceptiv Deliv Syst 1:371–377

Kurz KH, Meier-Oehlke P (1983) Jet-injection local anaesthesia for fitting and removal of IUDs. Contracept Deliv Syst 4:27–32

McKenzie R, Shaffer WL (1978) A safer method for paracervical block in therapeutic abortions. Am J Obstet Gynecol 130:317–320

Van Santen MR, Haspels AA (1981) Interception by postcoital IUD insertion, a review. Contracept Deliv Syst 2:189–200

Weström I, Bengtsson LP, Márdh P (1976) The risk of pelvic inflammatory disease in women using intrauterine contraceptive devices as compared to non-users. Lancet 2:221–224

Wolf AS, Krieger D (1980) Bedeutung der bakteriellen Kontamination von Intrauterinpessaren. In: Huber A (ed) Probleme der Kontrazeption bei den Jugendlichen. Exerpta Medica, Amsterdam, pp 124–132

Discussion XIII

Foster: You use 40 mg. This is a very small dose.

Van Santen: When I am carrying out a laparoscopic sterilization with local anaesthesia I always use the same solution. One or two drops on the top of the tube do also have an immediate effect on the peritoneum. But it does not seem to have the same good effect on the uterus.

158

Frontal EEG/EMG Analysis:
A Method of Assessing Depth of Anaesthesia.
First Experience with an "Anaesthesia and
Brain Activity Monitor"

W. Rating

Much is known about the influence of anesthetics on the electrical activity of the central nervous system (electroencephalogram, cerebral function monitor, Fourieranalysis) and there are also many reports of the effects of anaesthetics and muscle relaxants on the electrical activity of the frontalis muscle. The interpretation of this phenomenon during anaesthesia is well described but these procedures are too complicated (and also rather expensive) for the functional assessment of the central nervous system during routine anesthesia.

An "Anaesthesia and Brain activity Monitor" (ABM) has recently been developed by Datex instrumentarium OY (Helsinki, Finland) that can easily be applied during anesthesia. By means of computerized frequency analysis and using the same surface electrodes, this instrument can separately register the electrical activity from both the frontalis muscle (EMG) and the cerebral cortex (EEG). In the case of the latter, a one hundred per cent reading is given between O and 50 µV in the 0–20 Hz frequency range using zero crossing frequency techniques, and in the case of the EMG, between O and 15 µV. In both cases there is a semi logarithmic display. In addition the monitor can assess neuromuscular block and can measure carbon dioxide concentration in the expired gases.

The monitor was routinely applied to 100 patients in whom general anaesthesia was induced by a number of anaesthetic agents. In view of the fact that, in conscious patients, electrical activity of the frontalis muscle causes severe interference during electroencephalography, EEG changes during induction and recovery are not discussed.

An obvious application of the ABM is in the early detection of accidents or near accidents during anesthesia and surgery. For instance, cerebral depression as a result of overdose of anesthetics can be easily detected as can the immediate loss of cerebral activity that is seen after circulatory arrest. Recovery following succesful resuscitation is equally easy to follow. The monitor provides an immediate check of the patient's condition should the cable to the electrocardiogram become disconected. Apart from the above, stable EMG and EEG activity during the course of anesthesia, is of great assistance to the anaesthetist.

Frontal Electromyographic Activity (EMG) is influenced by psychomotoric reactivity and this, during anesthesia, is dependent on the degree of hypnosis, analgesia, and central muscle tone as well as the strength of surgical stimuli. Muscle tone is mainly influenced by muscle relaxants and to a greater or lesser extent by anesthetics.

Electromyographic Activity During Induction of Anaesthesia

When induction of anaesthesia is performed using intravenous agents such as etomidate (0.15–0.2 mg kg^{-1}) and methohexitone (1–1.5 mg kg^{-1}) only minor EMG depression is observed. This may be due to the occurrence of the abnormal muscular activity that is often seen after both agents. Alternatively it may be due to the placing of the anaesthetic mask on the face and the administration of ventilatory support. The same picture is seen when fentanyl is administered in intravenous doses of up to 0.3–0.4 mg.

Thiopentone, flunitrazepam, and midazolam, on the other hand, are followed by a more pronounced depression of the electrical activity of the frontalis muscle as is also seen following hypnotic and analgesic doses of ketamine. The same effects are seen when low doses of fentanyl (0.1–0.2 mg) are administered together with sedatives. This does not change when the patient receives ventilatory support and it may be that EMG depression after sedatives together with analgesics may be a good indicator of sufficient depth of anaesthesia.

Nitrous oxide in concentrations up to 70 volumes per cent causes no clear change in the EMG. In some cases a slow induction was performed with 70% nitrous oxide and enflurane or fluothane. The concentration of the agents was increased in steps of 0.5% every 1 or 2 min in a semiopen anaesthetic system. Enflurane concentration was increased to 4% and fluothane to 3%. Within 15 to 20 min a decrease in muscle-activity to less than 10 is seen (1–1.5 μV). This does not imply a loss of reactivity, because even at this low level of EMG activity motor reactions can still be observed following laryngoscope introduction or endotracheal intubation.

Supramaximal doses of muscle relaxants very rapidly reduce the frontal EMG activity. When small doses of relaxant are given after a sedative or anxiolitic drug, the rate of decline of EMG activity is accelerated. If relaxants are administered together with analgesics, the point of the maximum decrease of the EMG will indicate the earliest moment that acceptable intubation conditions will be present. This may occur well before complete peripheral loss of neuromuscular transmission is detectable.

A general conclusion is, that during induction of anaesthesia, EMG baseline is dependant on muscle relaxation. However, myographic depression following hypnotics administered together with analgesics is a more reliable indicator of sufficient depth of anaesthesia than is that caused by inhalation anaesthetic agents.

During the maintenance of general anaesthesia, both the level and the rate of change of the EMG signal give an indication of the balance between surgical stimulation and the depth of anaesthesia. The overall reactivity depends on the anaesthetic agent that is employed. For instance, nitrous oxide with intermittent fentanyl, results in large and rapid changes in EMG activity following surgical stimuli if the anaesthesia is too light. If fentanyl is then administered in sufficient dosage activity decreases relatively rapidly. If muscular relaxation is inadequate, sudden increase in EMG is normally followed immediately by forceful motor reactions. Occasionally a small dose of fentanyl may be so effective that only slight changes are seen and these disappear spontaneously.

Our experience with different combinations of analgesics and hypnotics or sedatives lead us to suspect that hypnotics and sedatives have their greatest influence on the EMG baseline activity, while analgesics effect the speed of change in the reading. For instance, small persistent increases in EMG activity during nitrous oxide-fentanyl anaesthesia can, in the absence of indications of lack of analgesia, only be reduced by hypnotics and sedatives (and of course by muscle relaxants), but seldom by additional doses of fentanyl. On the other hand, hypnotics and sedatives do not prevent sudden change in EMG-level, though they may delay the onset of movements. Similar effects are produced during infusion of etomidate and low doses of alfentanil. Increasing the dose of alfentanil will tend to damp the EMG reactions and if the dose of etomidate is reduced at this moment EMG baseline will gradually rise. Under anaesthesia with nitrous oxide and less than 1 MAC of enflurane or halothane relatively rapid increases in EMG activity can be seen, while above 1 MAC much smaller slower EMG changes are observed.

Muscle Relaxants

These have a large influence on the EMG-baseline. In particular this is seen when anaesthetic agents are used that have little effect on central muscle tone. Small doses of muscle relaxants have little influence on the acute pain reactions that are seen in the EMG. During continuous infusion of intravenous anaesthetic agents, small dosages of muscle relaxants can reverse these changes.

It is interesting to study how effectively frontal EMG may be used for the assessment of the adequacy of surgical relaxation and to make a comparison with standard neuromuscular blockade monitoring. If the first response of a "train of four" stimulus (the T1) is between 3%–5% of control EMG gives no useful information. If T1 is in the region of 5% or 6% of control, which might be seen when intravenous agents were used that had little effect on central muscle tone, increase of EMG activity may sometimes give an indication of insufficient surgical relaxation, though more reliable results were obtained when T1 was between 5% and 20%. When anaesthesia was administered using agents that produce relaxation by central depression, such as enflurane or fluothane, increase of T1 up to 30%–35% of control was to be observed before surgical relaxation became insufficient.

Electroencephalogram

The ABM system may misinterpret high frequency, low amplitude EEG patterns, such as may be seen under intermittent fentanyl anaesthesia, and may present it as a low frequency, low amplitude signal. It would appear that zero crossing frequency techniques reach a limit with high frequency, low amplitude EEG. Only when hypnotics or sedatives are added and the cortex activity changes to theta and delta, a good correlation is to be seen. This is the reason why no discussion is presented on the influence of intravenous agents on the EEG.

Inhalation Agents

Nitrous Oxide

Normally after about 5 min of nitrous oxide administration, a phase of excitement is seen that interferes with the EMG recording. Up to this moment no specific changes were observed.

Enflurane

When this agent is administered with 70% nitrous oxide in oxygen in increasing concentration a temporary increase of EEG frequency from 8–10 Hz to 10–14 Hz, in the presence of decreasing amplitude can be seen. Subsequent amplitude increases varied from patient to patient. Up to 1.5% enflurane, the frequency decreases more than at higher concentrations. Beyond 2.5%–3% "burst suppression" phenomenon becomes likely. Below 7–8 Hz, which was usually seen at an enflurane concentration of 1.3%–1.5%, pain reactions were not observed. In the individual patient the EEG-frequency correlates well with the enflurane concentration during stable anaesthesia. Some patients have a tendency to develop hypotension after minor or moderate doses of enflurane and they develop a low amplitude pattern at a frequency of 5–6 Hz or less. Under these conditions, frequency is not a good indicator of sufficient depth of anaesthesia. It is not clear whether, as we suspect, the zero crossing frequency technique is unsuitable under these conditions, or if other factors such as hypotension may be causing this "cerebral depression". It should not be forgotten that under intermittant fentanyl-nitrous oxide anaesthesia and during shock comparable depressed EEG-patterns were also seen.

Fluothane

This agent, in combination with 70% nitrous oxide in oxygen causes similar EEG changes to enflurane. The frequency does however seem depressed to a greater extent and there is more amplitude variation than is the case with enflurane. At fluothane concentrations below 0.7%, strong surgical stimuli, such as mesenteric or peritoneal traction may cause a severe fall of frequency with an increase in amplitude. This reverts to the original pattern after removal of the stimulus. If the fluothane concentration in this case is suddenly increased, then it is difficult to correlate EEG depression with surgical stimuli or sufficient depth of anaesthesia until a new steady state is achieved. Hence an unstable course of anaesthesia can provoke a chaotic picture of electroencephalographic activity.

Low-flow-technique and Closed Circle Anaesthesia

When the above techniques are employed, both enflurane and fluothane produce a much more stable electroencephalographic picture than is seen when semi-closed or even semi-open systems are employed.

Pneumatic Controlled Circulation

W. L. den Dunnen and T. Mostert

Introduction

We know that the alveoli and the capillaries around the alveoli play an important role in gas exchange. It is also known that alveolar and capillary compliance contribute a major part to the total lung compliance. If these structures have a great compliance it must be possible to exert an influence on the alveolar gas compartment (inflating of ventilatory gases), which has an immediate effect on the capillary blood circulation. On the other hand, if the capillaries around the alveoli (the capillary meshwork) have a great compliance they must exert an influence on the alveolar gas compartment during right ventricular contraction, which fills the capillaries with blood, increasing the total interalveolar volume. The increase in interalveolar blood volume should lead to displacement of alveolar gases.

This natural influence can be easily measured in patients during prolonged apnoeic periods. Every anaesthetist recognizes the movements of the anaesthetic balloon of a circle system synchronous with the patient's heart beat. Also, cardiogenic oscillations are known from capnographic studies, which will be discussed later.

However, some clinical findings show that the "hydraulic" system of the lungs exerts an influence on the "pneumatic" system. If we want to exert the opposite effect we must insufflate the ventilatory gases in such a way that the alveolar gas pressure goes low during right ventricular contraction and that the alveolar gas pressure rises during right ventricular diastole. If we can perform artificial ventilation in such a way, it will employ the natural movements of the alveolar septa, already caused by right ventricular influence.

The insufflation of gases has to be adjusted synchronously with the pressure in the capillaries, providing a maximum difference between pneumatic and hydraulic pressure in the alveoli. In our technique the pneumatic control is adjusted in such a way that the momentary blood pressure in the capillary meshwork determines O_2 and CO_2 diffusion. The ventilation is adjusted in such a way that the requested specific gas exchange corresponds with the momentary transmembranous pressure. This method of ventilation will guarantee the most efficient gas exchange.

During spontaneous breathing inspiration causes subatmospheric intrathoracic pressure, which helps blood to flow back to the heart. During expiration intrathoracic pressure rises, which causes reduction in suction of venous blood. Cardiac output (left-sided) rises a little because of compression of the lungs. However this interaction of pneumatics and hydraulics in the thoracic cavity does not lead to a real change in cardiac output, although an obvious interaction is present.

It is important that the interaction of the pneumatic and the hydraulic systems has no negative effect. During exercise the statistical occurrence of interaction between hydraulics and pneumatics increases. An optimal synchronization is necessary to guarantee most effective breathing and circulation at the lowest energy level. It also makes sense to suppose that a natural synchronization between circulation and ventilation is present at rest – and especially during a period of sleep or anaesthesia. This phenomenon will be discussed later.

Intermittent positive-pressure ventilation (IPPV) causes a disturbance of the natural relationship between respiration and circulation in the thoracic cavity because of the continuous mean positive pressure. The interaction between circulation and respiration is predictable: it is directly related to the ventilation frequency. IPPV is applied with low dP/dt of the insufflated gases. Due to the low dP/dt the thorax is enlarged by the insufflated volume, causing acceptable interaction with the circulation. However, if positive end-expiratory pressure (PEEP) is applied, the interaction with the circulation might lead to serious problems. Venous return can be dramatically reduced, which causes changes in cardiac output. These changes in cardiac output will be compensated by autoregulation mechanisms to maintain a circulation as optimal as possible under the given circumstances. Due to the low dP/dt of the ventilatory gases and the static character of PEEP application, it appears to be impossible to obstruct the pulmonary blood flow. (The thorax will distend and right ventricular stroke work will increase.)

During IPPV with or without PEEP, one can observe a slight alternated increase in pulmonary and aortic pressure when the gases inflate the lungs. This clinical finding and the hypothesis of natural interaction and synchronization between pneumatics and hydraulics made us decide to synchronize the heart duty cycle with the respiratory cycle. We thought it would be possible to control the hydraulic system by insufflating the gases in such a way that alveolar gas pressure wil be low when the lung capillaries and veins are filled with blood from the right ventricle and that the alveolar gas pressure will be relatively high when the ventricular contraction has ended.

However, as soon as the alveoli are inflated more than is normal during spontaneous breathing, the energy of the insufflated gases will immediately be transferred to the lung capillaries and veins, causing acceleration of blood to the left atrium. This externally controlled lung pneumatics influences lung hydraulics, which causes a lower vascular resistance in the lungs. It causes a smaller "residual" blood volume in the capillaries and veins. When right ventricular systole starts, the pulmonary end-diastolic pressure will be reduced, so right ventricular stroke volume will increase, although right ventricular stroke work is reduced. To perform the type of ventilation outlined here requires a special ventilator. An adequate tidal volume needs to be insufflated heart beat by heart beat. The second criterion for the performance of this kind of ventilation is that the insufflation should be performed with high dP/dt of the ventilatory gases.

However, a jet insufflation technique is very suitable for the performance of this kind of ventilation. A special research ventilator (Pulmonary Pneumocontroller, Mostert BV, Physical Laboratory, Emmeloord, Holland) was designed to perform this kind of heart-beat-synchronized jet ventilation. Why the application of a jet technique? The answer is easy. Due to the impedance characteristics of the airway, it is possible to insufflate a relatively small tidal volume into the trachea at relatively low intratracheal peak pressure, causing a relatively high mean pressure more distally in the airway. A jet technique may provide an abnormally high pressure as the alveoli are approached. The mean pressure in the bronchioli may be even higher than the intratracheal mean pressure at the same time! This physical property is comparable with the vascular system, showing higher peak pressures in the smaller arteries than in the aorta.

Insufflation of ventilatory gases with low dP/dt (IPPV) will show the highest gas pressure in the trachea. The further the lungs are penetrated, the lower gas pressure that will be measured (open trumpet model). On the other hand, if we insufflate the ventilatory gases with high dP/dt at relatively high frequencies (heart frequency or higher), we have to consider the impedance characteristics of the airway. If we want a certain amount of gas to pass through a narrow tube, it will be more difficult than making it pass through a wide opening. If we want to perform this in a very short time it can easily be understood that the gas pressure will rise! The "open trumpet model" cannot be used when applying a jet technique.

If we ventilate in synchrony with the heart beat, we can insufflate at different moments of the heart duty cycle. Triggering should be performed in such a way that intra-alveolar gas pressure and peri-alveolar capillary blood pressure are fully counterphased with a certain phase-shifting. We called this new technique "pneumatic controlled circulation" (PCC).

Methods

Insufflation of the gases in the trachea is performed through a 10-Charrière suction catheter with one opening on the tip only (Unoplast, Denmark). The catheter is placed in a routine endotracheal tube (9 mm internal diameter). The tip lies about half-way down the tube. For monitoring endotracheal pressure and temperature, a 7-gauge Swan-Ganz thermodilution (TD) catheter is used. The catheter is modified by shortening the tip just distally to the thermistor and is positioned about 2 cm above the carina with the temperature sensor just below the tip of the endotracheal tube.

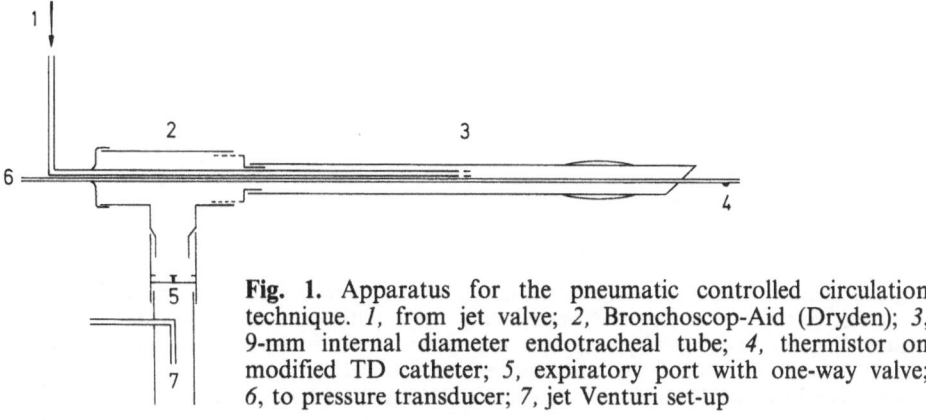

Fig. 1. Apparatus for the pneumatic controlled circulation technique. *1*, from jet valve; *2*, Bronchoscop-Aid (Dryden); *3*, 9-mm internal diameter endotracheal tube; *4*, thermistor on modified TD catheter; *5*, expiratory port with one-way valve; *6*, to pressure transducer; *7*, jet Venturi set-up

The proximal and insufflation ports of the TD catheter are permanently closed. The insufflation and Swan-Ganz catheters are fixed into a special T-piece (Bronchoscop-Aid, Dryden, USA). The expiratory port is closed with a one-way valve to prevent entrainment of room air during insufflation, and minute volume is monitored through this port using a Wright respirometer. On the expiratory port a special suction device can be mounted, using a second jet valve, providing faster expiration. The suction power operates on the Venturi principle and can be accurately adjusted. The apparatus is shown in Fig. 1.

The Ventilator

The ventilator will be discussed on the basis of the functional diagram shown in Fig. 2. *1* is the gas mixer. An emergency bypass button provides immediate 100% oxygen supply. *2* is the gas-mixing chamber. On the top of this chamber a rotameter is mounted. Here a Water's set or other anaesthetic device can be connected. Oxygen concentration control is performed with an external oxygen concentration monitor, sampling from the mixing chamber. *3* is the adjustable jet-pressure-reducing valve with manometer. The maximal jet pressure which can be set depends on the setting of the gas-mixing valves. However, in most hospitals the central gas supply guarantees at least 4–6 bars on the pipeline system, which is more than enough to set the requested jet pressure.

4 is the humidifier and heater. This is a hermetically closed acrylic container, partly filled with sterile distilled water. The gases are introduced via a gas dispenser into the bottom of this chamber. At the bottom a regulated heating device is mounted which is controlled by a feedback mechanism from a temperature sensor mounted at the top of the chamber. The gases are passed via a consideration trap into a specially designed jet valve. The heater can be adjusted to guarantee the humidified jet gases being supplied at body temperature. *5* is the specially designed jet valve. It is placed close to the patient. Placing this valve near the patient reduces the tubing dead space considerably. The actual dead space, including the 10-Charrière catheter, is less than 3.5 ml in total. The jet valve operates in such a way that it is always in the closed position in case of electronic failure. The maximum opening time of the jet valve is the insufflation time (in milliseconds) which has been set. This prevents the risk of prolonged insufflation during inadequate ECG inputs.

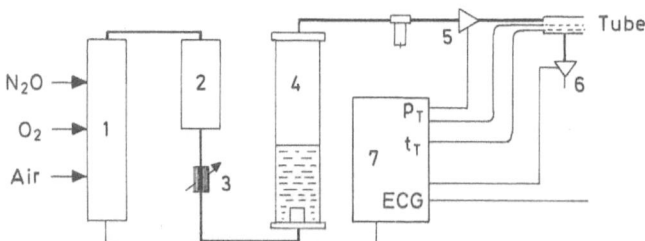

Fig. 2. Functional diagram of the pulmonary pneumocontroller. *1*, gas mixer; *2*, gas-mixing chamber; *3*, adjustable jet-pressure-reducing valve with manometer; *4*, humidifier and heater; *5*, specially designed jet valve; *6*, expiratory Venturi set-up which can be mounted on the special T-piece of the endotracheal tube; *7*, electronic function controller: p_T, patient's endotracheal pressure; t_T, patient's endotracheal temperature

6 is the expiratory Venturi set-up, which can be mounted on the special T-piece of the endotracheal tube. It operates with a jet valve the same as that outlined at 5. This valve can be used to perform faster emptying of the airway tract. It operates in counterphase with the patient's jet valve. It has an "auto-shut-off" after 1 s, preventing continuous suction of the airway in the case of failing ECG input. 7 is the electronic function controller. Inputs are the patient's ECG, endotracheal temperature and endotracheal pressure. The R-R interval of the ECG is continuously displayed in milliseconds. The endotracheal temperature is displayed in degrees Celsius. The endotracheal pressure is displayed on a ventrical scale on which light-emitting diodes show the pressure fluctuations. Besides this, the endotracheal pressure is digitally displayed. To increase the patient's safety, a maximum and minimum endotracheal pressure limit can be set. If the maximum is reached the ventilator switches to expiration mode. If the minimum pressure is reached (if the Venturi device is used) the ventilator stops until the next insufflation. Also the pressure fluctuations are continuously monitored. If the fluctuations are too low the ventilator starts an audible alarm. In our opinion this way of monitoring the endotracheal pressure provides excellent patient security if a jet ventilator is used.

The electronic settings which can be made with this ventilator are:

Heart-synchronous mode. Insufflation pulses from about 20 ms to 1 s can be set. A delay can be set using the patient's ECG as a time reference. The jet valve can be triggered by an external trigger signal or as a function derived from the pulmonary artery pressure and the ECG. During the experiments we preferred a digital hand setting, although we have developed microprocessor programmes to calculate the control function.

Heart-asynchronous mode. In this mode the ventilator operates as a normal jet ventilator, as described by other authors (Sjöstrand 1977).

Anaesthetic Technique

All patients were given oral premedication – 2.5 mg lorazepam (Temesta) – about 2 h before induction. They were given an intramuscular dose of 5 mg dehydrobenzperidol (droperidol) and 0.1 mg fentanyl 45 min before induction. Normovolaemia was ensured by pre-operative infusion of Ringer's lactate solution. Anaesthesia was induced with etomidate (Hypnomidate 0.2 mg/kg body wt.), and after muscle relaxation had been produced by a dose of suxamethonium (1 mg/kg body wt.) the patients were intubated with a disposable Vygon endotracheal tube (9 mm internal diameter). Anaesthesia was maintained with 33% oxygen and 66% nitrous oxide, supplemented with neuroleptanaesthesia using dehydrobenzperidol and intermittent doses of fentanyl. Muscular relaxation was obtained using alcuronium (0.2 mg/kg body wt.), and the patients were ventilated with a Dräger Spiromat ventilator to achieve an end-expiratory carbon dioxide level between 3 and 4 vol.%.

Patients in whom ECG and capnographic studies were done received the same premedication. Anaesthesia was induced with thiopentone (3 mg/kg body wt.). Intubation was facilitated with suxamethonium (1 mg/kg body wt.). Anaesthesia was maintained with 33% oxygen and 66% nitrous oxide, together with enflurane

(Ethrane) between 1.5 and 2.5 vol.%. These patients were breathing spontaneously on a one-way anaesthesia circuit.

The PCC ventilation research started after normalization of the circulatory volume. Infusion therapy was continued until pre-operative blood pressure was reobtained. During steady-state anaesthesia short runs were made with the jet valve ventilator using the same oxygen concentration of the ventilatory gases. All the patients selected (six) were undergoing operative procedures with minimal blood loss, such as peripheral vascular surgery.

Monitoring

Intra-arterial pressure monitoring was performed in all patients, using an 18-gauge intra-arterial catheter, introduced into the radial artery. A 7-gauge Edwards Swan-Ganz thermodilution catheter was inserted using a Cordis catheter introducer system via the cubital vein. A 14-French "National Catheter" gastric tube with an oesophagus balloon was inserted. The stomach was emptied and the gastric tube was closed off. This enabled monitoring of the intra-oesophageal pressure to be performed via the oesophageal balloon. The intravascular catheters and the balloon of the oesophagus catheter were connected to Gould pressure transducers (type P231D) and the intratracheal pressure Swan-Ganz catheter was connected to an Ailtech microtransducer (type MS20EA44 ABS). A Siemens capnograph was used during IPPV. ECG and pressure monitoring was performed with Hewlett-Packard devices (78000 series). A four-channel MFE strip-chart recorder was used (type 1420) with an eight-channel recorder (type Gould Brush 481).

For cardiac output measurements a WTI cardiac output monitor was used. Oxygen concentration monitoring was effected with a Bio-Marine analyser (type 202A). To study natural cardiopulmonary synchronization a Dräger capnograph (Capnolog) was used with the cuvette system immediately attached to the endotracheal tube.

Results

Notice the oscillations in the capnogram (Fig. 3) during a period of prolonged apnoea. These oscillations are known as "cardiogenic oscillations". Nunn and Hill (1960) found that the CO_2 rose sharply with each heart beat, at the carina, in the apnoeic patient, and alveolar gas is known to reach the glottis under these conditions. These oscillations are ascribed to the reduction in anatomical dead space during hypoventilation as a result of the heart beat providing some degree of ventilation (Smalhout and Kalenda, p. 134). Elam and Brown (1955) recorded the volume of the displaced gases as varying between 3 ml and 48 ml. Notice the change in amplitude of the oscillations, which will be discussed later. Smalhout and Kalenda (p. 141) summarize the factors concerned in the capnographic appearance of cardiogenic oscillations as follows:

1. The presence of negative intrathoracic pressure
2. A pulse/respiration ratio of approximately 5.4/1 in an adult and $\pm 4/1$ in a child

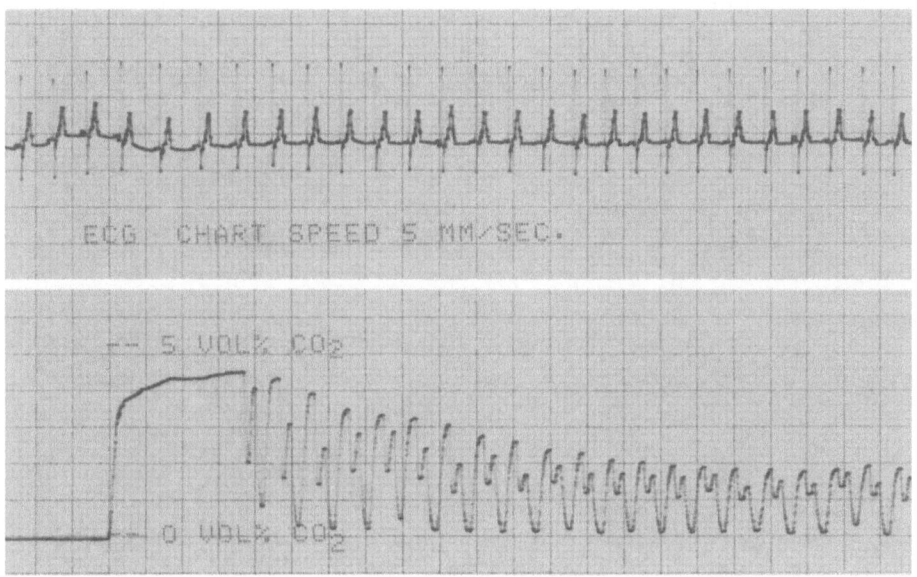

Fig. 3. Simultaneous capnogram and ECG recording

3. A diminution in the vital capacity/size of heart ratio
4. A low inspiration/expiration ratio
5. A prolongation of the expiratory phase
6. Apnoea
7. A very low tidal volume
8. Muscular relaxation.

From the beginning of the experiments onwards there always existed considerable interaction between the first derivative of the intratracheal pressure (dP/dt) and the blood circulation. It must be emphasized that relatively long insufflation times were used compared to those of other authors (Jonzon et al. 1971; Heijman et al. 1972; Sjöstrand and Erikson 1980).

Figure 4 shows tracings obtained from two different experimental runs of ventilation with the pneumocontroller. The simultaneous registration of oesophageal pressure (Fig. 4b), pulmonary artery pressure (Fig. 4c), intratracheal pressure (Fig. 4d), and central venous pressure (Fig. 4e) is demonstrated as closely related to the patient's ECG cycle (Fig. 4a). Insufflation times can be derived from tracing d in Fig. 4. In Fig. 4 the two experiments are superimposed to show the effects of two different ventilator settings. In tracing 4a the ECG tracing of experiment I is shown only, although there existed little change in ECG frequency. The dotted lines of experiment II are drawings derived from the original tracks. One should concentrate on the first three heart duty cycles because of phase-shifting.

However, two different experiments are registered (solid line and dotted line). The jet ventilation (Fig. 4d) is set to inflate at two different moments (expiration delay 0 and 700 ms from the top of the QRS complex). In experiment I (solid line, Fig. 4c) a setting is shown in which peak intratracheal pressure (Fig. 4d, solid line) coincides with ventricular systole. An important difference can be seen in the

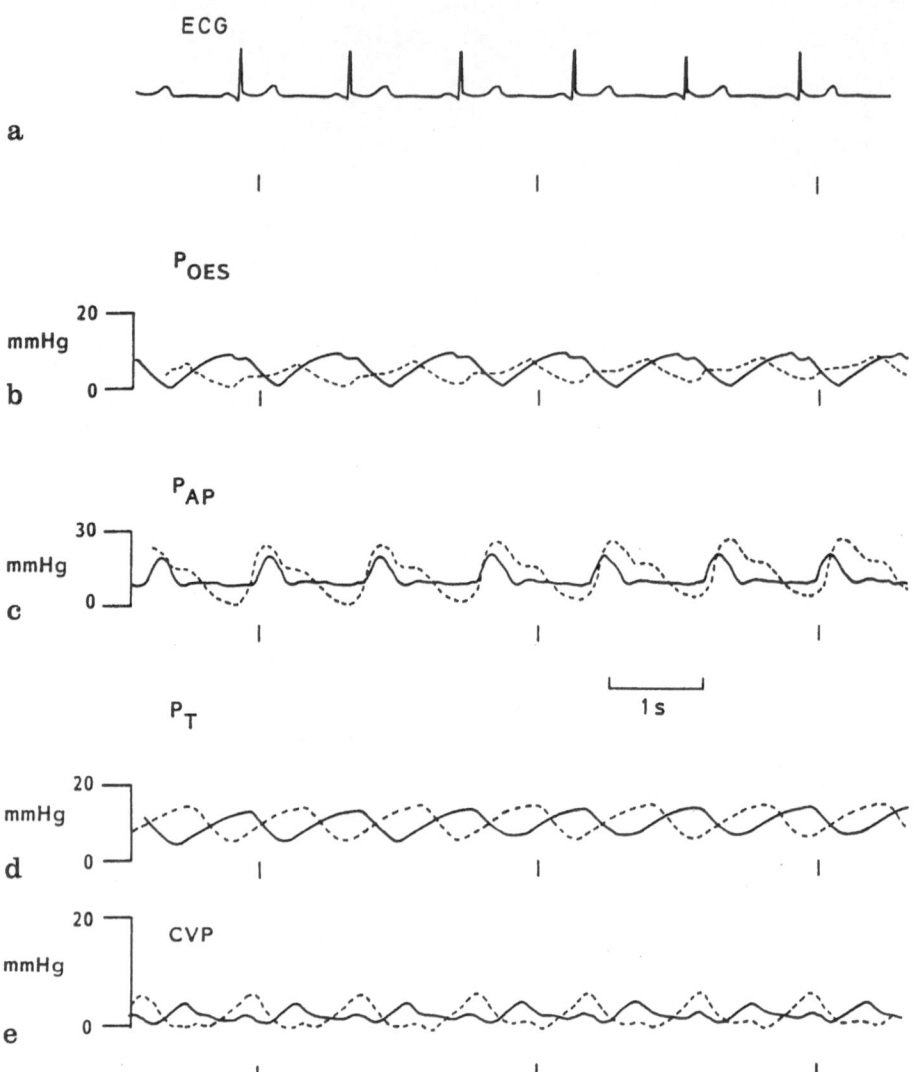

Fig. 4a–e. Registration of experiments I and II. P_{OES}, oesophageal pressure; P_{AP}, pulmonary artery pressure; P_T, intratracheal pressure; *CVP*, central venous pressure

effect on the pulmonary artery pressure (Fig. 4c). It can be seen that the pulmonary artery systole (Fig. 4c, solid line) is reduced. The dicrotic notch is less pronounced and the pressure remains almost constant after closure of the pulmonary artery valve until the next right ventricular systole. The corresponding pressure tracings (Fig. 4b, d, solid lines) show pressure changes in phase with the intratracheal pressure (harmonic resonation).

Figure 4c, d shows the association between hydraulic pressures and the insufflation pressure, and suggests great changes in transthoracic pressures (Fig. 4b). The small dip in the top of the oesophageal pressure wave is related to the heart contraction.

In experiment II, the tracings of which are indicated by dotted lines, the beginning of the expiratory phase of the ventilator is triggered to occur at a point

700 ms after the top of the ECG-QRS complex, so peak airway pressure (Fig. 4d, dotted line) occurs, in this instance, at a point 400 ms before the next QRS complex, i.e. during ventricular diastole. In this case the pulmonary artery pressure (Fig. 4c, dotted line) shows higher systolic peak levels, the dicrotic notch looks more physiological and pressure continues to decline during diastole to reach much lower levels than was the case in experiment I (solid line). The great difference in area under the two pulmonary artery curves (Fig. 4c), which suggests a quantitative rise in cardiac output during the second setting (dotted line), should be noted. The oesophageal pressure shows smaller pressure variations, suggesting a smaller change in transthoracic pressure (Fig. 4b, dotted line, experiment II) under these conditions.

The central venous pressure traces vary with the two settings (Fig. 4e). In experiment II (dotted line) the atrial contraction wave is very marked, indicating greater ventricular filling, and hence implies a larger right ventricular stroke volume. This is reflected by the greater area under the pulmonary artery pressure curve (Fig. 4c, dotted line). Also an increase in the X-dip amplitude indicates an increase in right ventricular stroke volume.

Although aortic pressure is not shown, we want to emphasize that there was not much influence in the short term on the arterial pressure at the two settings, showing that the left ventricular contraction force remained primarily unaffected, but a significant decrease has been found during longer periods (after 3–4 min) of ventilation at setting I (Fig. 4c, solid line). The runs were too short to allow evaluation of the effect on the body circulation. Later on we repeated the same research in animal experiments (sheep). For ethical reasons we did not go on with this research in patients because of the risks involved before all the factors were analysed.

Figure 5 shows a ventilator setting at which the insufflation rate does not totally equal the heart rate as compared in experiments I and II (heart rate = 68 beats/min and jet insufflation rate = 60 jets/min). An obvious influence on the

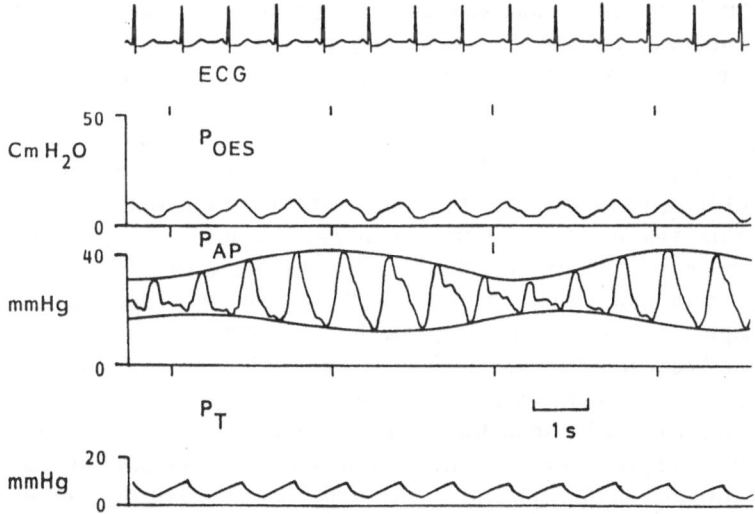

Fig. 5. Interference with pulmonary artery pressure caused by asynchronous high-frequency positive-pressure ventilation. P_{OES}, oesophageal pressure; P_{AP}, pulmonary artery pressure; P_T, intratracheal pressure

pulmonary artery pressure is shown with this changing timed insufflation during the heart duty cycle. The pulmonary artery pressure shows regular variation in amplitude, depending on the degree of phase or counterphase of the ventilation with respect to the heart beat. When peak airway pressure coincides with ventricular systole, the pulmonary artery pulse pressure is at its minimum, and the diastolic pressure is at its maximum. On the other hand, when the peak airway pressure is at its maximum during diastole, the pulmonary artery pressure is at its maximum, and the diastolic pressure is at its minimum.

One has to realize that this phenomenon occurs only with intratracheal pressures between 5 and 15 cmH$_2$O. (The harmonics from this base insufflation frequency also show this interference.) Notice the relatively long periods of increased amplitudes of the pulmonary artery wave curve compared to the relatively short periods of decreased amplitude. The harmonic oscillations as shown here disappear if the insufflation is completely synchronized with the heart rate. If this occurs it can never be predicted in which phase of the heart cycle synchronization will be obtained, unless pulmonary artery pressure monitoring is performed. Also oesophageal pressure should be monitored. Harmonic resonance of the intratracheal pressure in the oesophageal pressure indicates a high transthoracic pressure, which implies a very dangerous jet ventilation setting.

Discussion

Although our investigations are not yet complete we want to show the possibility and the value of utilizing the naturally existing cardiopulmonary interaction. We want to demonstrate that jet insufflation techniques can seriously interfere with the pulmonary circulation, although this has not been found by other authors (most recent publication: Otto et al. 1983). This interference can be used to improve the pulmonary blood circulation and thus the total body circulation. This has also been suggested by other authors (Dedhia 1981; Klain and Smith 1977; Klain and Keszler 1980). To understand this PCC we have to reconsider our knowledge of pulmonary physiology. Pulmonary pneumatic influence on the blood circulation is a natural interaction occurring during spontaneous breathing.

Any physiology textbook will explain airway resistance in the following terms: The contribution to resistance is directly related to the total cross-sectional area of each division of the airway system and not the diameter of any one given airway within that subdivision. The total cross-sectional area increases as the airways become more peripheral, and resistance decreases. We wish to emphasize that this explanation of airway resistance is basically not true, although it can be used as long as displacement of ventilatory gases with static pressure changes (low dP/dt) is involved. According to Poiseuille's law, the absolute airway resistance increases with entry into the airway system, due to the decrease in diameter of the channels. Gas molecules decelerate as they enter the lungs (there is higher resistance as the alveoli are approached, and there is therefore a lower gas flow and a higher cross-sectional volume, causing a lower gas pressure). During spontaneous breathing the interpleural pressure changes from about -4 mmHg to -9 mmHg. The energy provided by the respiratory muscles is not sufficient to ensure adequate suction of the ventilatory gases into the alveoli. Inspiratory movements cause the lungs to be filled with air, but the alveoli are hardly ventilated by this. It should be realized that the airway beside the alveoli has a high compliance, so the greatest part of the inspiratory volume remains in the airway and does not enter the alveoli directly. It will be absorbed by the cardioalveolar respiration pump, as explained under Conclusions. In our opinion, "cardiohydraulic" influence on the ventilatory system is a natural interaction, occurring continuously. To explain this it is necessary to reconsider our ideas about circulation in the lungs.

During right ventricular systole, the stroke volume enters the lung circulation meshwork. The capillary system around the alveoli has a higher impedance than the pulmonary artery, so in the narrowing vessels the potential energy of the blood is converted into kinetic energy, leading to an increase in the capillary diameter. The interalveolar blood volume is increased, so alveolar gases are displaced. On the basis of our own research we are convinced that the alveoli are partly emptied by this mechanism. There is an immediate relationship between right ventricular stroke volume and "total alveolar stroke volume". This means that the alveoli are ventilated by energy from the right ventricle. Every systole the alveoli are partly emptied, and every diastole fresh air is "inhaled" through the bronchiolar ducts. Alveolar gas exchange is very dynamic and is not refreshed breath by breath but heart beat by heart beat.

Cardiogenic oscillations as shown in Fig. 5 fit with our ideas about alveolar gas transport. The decrease in amplitude of these oscillations can be interpreted as follows. In apnoeic periods the displacement of the ventilatory gases is only performed by the heart, comprising two components:

1. Right ventricular stroke volume entering the pulmonary capillaries, causing an expiratory component
2. The portion of the left ventricular stroke volume leaving the thoracic cavity, causing an inspiratory component.

To give an idea of the quantity of these components we have to consider the capnogram during apnoeic periods. We can see CO_2 being expired synchronously with the ventricular systole, implying a positive algebraic sum of these components, which means that the expiratory component due to the right ventricular stroke volume is larger than the inspiratory component, due to the decrease in volume of the heart during systole.

In apnoeic periods the total interalveolar blood volume increases (tendency of positive outflow of the alveoli), so blood capillary resistance is decreased. The alveoli remain in "expiration position", which will alter the algebraic sum. Also, near the capnograph sensor room air is entrained every heart beat, so the concentration of CO_2 being measured is reduced.

In all six patients studied, a greater or lesser degree of interaction between the ventilator and the pulmonary vascular state of the subject was observed. The direction and degree of this interaction was determined in all cases by the degree of phase or counterphase of the two systems. It should be stressed that these effects were observed at modest intratracheal pressures (5–15 cmH$_2$O) that were well within the accepted range of pressures used in other ventilatory patterns. The findings in all six patients revealed that, when peak (alveolar) inspiratory pressure coincided with ventricular systole, severe decreases (of up to 50%) in right heart output were produced.

It appears that it is possible, by ensuring correct relationship between jet insufflation triggering and the heart duty cycle, to increase right heart output and pulmonary artery pulse pressures.

In the meantime, sheep experiments have confirmed that it is possible to increase cardiac output to such an extent that the aortic pressure can be almost doubled. After a short period of PCC with increased transmembranous alveolar pressures these animals showed severe attacks of bradycardia. Bradycardia disappeared after restoration of IPPV. Blood gases were excellent during the attacks. Autopsy revealed rigid, oedematous, blood-filled lungs with pronounced parenchymal infiltration. This bad lung status has also been mentioned in patients being treated with a combination of high-frequency positive-pressure ventilation and PEEP for longer periods (Bjerager et al. 1977, pat. no. 2).

Conclusions

The results of this study indicate a systematic interaction between the respiratory system and the circulation. This interaction can be used by calculated insufflation pulses to control the circulation externally. In our opinion respiratory physiology should be modified. On the basis of our findings we subdivide lung ventilation into:

1. Exogenous respiration, the already known respiration which mainly ventilates the larger airways
2. Endogenous respiration, the until now unknown alveolar ventilation caused by right ventricular ejection. The influence on the alveolar ventilation caused by the right heart is known as the "cardioalveolar respiration pump" (CARP).

Our physiological lung model provides a better understanding of alveolar gas transport. Exogenous respiration can be performed by thoracic excursions (natural breathing) or by artificial ventilation. Endogenous respiration is performed by the CARP and can be supported by synchronous exogenous respiration (PCC) or natural cardiorespiratory synchronization. All other ways of ventilation interfere with the CARP.

Animal studies have shown that during the application of this PCC technique all reserves of venous blood can be pumped to the left heart, causing an enormous temporary increase in cardiac output until the venous reservoir is nearly empty. This will occur when PCC is performed "in resonance with the CARP" (counterphased insufflation) at too high energy levels. This should be avoided because of the serious and irreversible damage to the lung tissue.

We are convinced that PCC must be applied in such a way that "venous emptying" does not take place, but that it has to be used to reinforce the naturally existing cardiopulmonary interference. If the application is performed in a proper way it can be used for many clinical problems such as right ventricular decompensation and pulmonary valve insufficiency, as a replacement for the intra-aortic balloon pump and for diagnostic purposes. It can also be used to ventilate patients at very modest intratracheal peak pressures (less than 10 cmH$_2$O), which do not interfere with the circulation at all. This CARP-resonance PCC technique reinforces the "endogenous" respiration. Due to the characteristics of jet ventilation the "exogenous" respiration is then unnecessary, so the thorax moves at minimal level.

As mentioned in the Introduction, it would make sense for a natural synchronization between ventilation and circulation to exist. One can observe incredibly accurate synchronization of the circulation and gas pump. Simultaneous registration of the patient's ECG and capnogram show – during sleep or anaesthesia (with spontaneous breathing) – that inspiration and expiration are fully synchronized. Very regularly, periods of inspiratory and expiratory phase-shifting can be observed. In our opinion this is very important to restore the left-right blood balance. Inspiration during right ventricular systole will increase the right ventricular stroke volume, while expiration in this phase of the heart contraction will have the opposite effect.

However, this very complex matter is still under clinical investigation. When our computer analysis is complete it will be published in a special paper concerning this subject. Now it can already be postulated that besides their respiratory function the lungs play an important role in controlling the circulation of the blood.

The results of this study indicate that, contrary to the assertions of other authors working with high-frequency ventilation and high-frequency positive-pressure ventilation, high-frequency ventilation may seriously influence the circulation at frequencies at or near the heart rate or at multiples thereof (harmonic frequencies). In our opinion jet ventilation techniques interfere mainly with the endogenous respiration under certain circumstances. It appears to be very difficult to ventilate animals and patients in phase with the ventricular systole. They all develop an increased heart rate and tend to oscillate around the dangerous setting. This causes prolonged periods of counterphase ventilation, leading unintentionally to increased alveolar transmembranous pressures. This will seriously damage the lung parenchyma, leading in the short term to oedema and damage of the capillary meshwork (acute haemorrhage). In the long term interstitial fibrosis might be the result, in spite of probably more favourable levels of blood gases at the moment.

References

Bjerager K, Sjöstrand U, Wattwil M (1977) Long term treatment with IPPV/PEEP and HFPPV/PEEP. Acta Anaesthesiol Scand (Suppl) 64:55–68

Dedhia HV (1981) Hemodynamic effect of high frequency ventilation in open heart surgery patients. Crit Care Med 9:163

Elam JO, Brown ES (1955) Carbon dioxide homeostasis during anaesthesia. 11. Total sampling for determination of dead space, alveolar ventilation and carbon dioxide output. Anaesthesiology 16:886

Heijman G, Heijman L, Jonzon A et al. (1972) High frequency positive pressure ventilation during anaesthesia and routine surgery in man. Acta Anaesthesiol Scand 16:176–187

Jonzon A, Öberg PA, Sedin G et al. (1971) High frequency positive pressure ventilation by endotracheal insufflation. Acta Anaesthesiol Scand (Suppl) 43:1

Klain M, Smith RB (1977) High frequency percutaneous transtracheal jet ventilation. Crit Care Med 5:280–287

Klain M, Keszler H (1980) Circulation assist by high frequency ventilation. Crit Care Med 7:232

Nunn JF, Hill DW (1960) Respiratory dead space and arterial to end-tidal CO_2 tension difference in anaesthetised man. J Appl Physiol 15:383

Otto CW, Quan SF et al. (1983) Hemodynamic effects of high frequency jet ventilation. Anesth Analg 62:298–304

Sjöstrand UH (1977) Pneumatic systems facilitating treatment of respiratory insufficiency with alternative use of IPPV/PEEP, HFPPV/PEEP, CPPB or CPAP. Acta Anaesthesiol Scand (Suppl) 64:123–147

Sjöstrand UH, Erikson IA (1980) High rates and low volumes in mechanical ventilation – not just a matter of ventilatory frequency. Anaesth Analg 59:567–768

Smalhout B, Kalenda Z (1975) Textbook of capnography, vol 1. Kerckebosch, Zeist, The Netherlands, pp 134–141

Discussion XIV

Healy: Some of the physiology which was presented I must completely disagree with. The "new physiology" and the explanations of the movements of the blood are wrong when judged against basic presentations in the standard works in our textbooks. It was said that the pulmonary circulation was a high-pressure system and it is upon that brick that the whole house stands. But I am afraid that has to be kicked from under the house. Everybody knows that the pulmonary circulation is a low-pressure system and a low-resistance system.

den Dunnen: This is the same problem found in articles about jet ventilation. I know that what I am saying is against the rules. But there is one thing about resistance: if the size of the airways decreases, the resistance must increase. A physicist told me that if you insufflate gas into the lung in a narrowing tract you get an increase in pressure. This happens because of the impedance characteristics of the airways. So who is making the mistake, the textbook, the physician or the physicists? It is very important to realize the impedance characteristics of the airways. In our sheep experiments we put a catheter deeper and deeper into the trachea during asynchronous jet ventilation. We saw a rise in pressure although the total resistance decreased. The diameter of the airways also decreases. The problem is that you cannot assume that all the airways are parallel. Therefore, the pressure behaves in a different way from that described in the medical textbooks.

If I apply a pressure fluctuation in the trachea, then the pressure and the resistance increase slightly, and the higher the frequency of the asynchronous ventilation the higher the pressure in the small airways is compared with the pressure in the trachea. And this is the same as what happens in the small arteries of the left-side circulation. Why is it true in the blood system and not in the gas system? It is the same physical law! There is no physiologist who can prove our results wrong.

McIntyre: I would like to remind the audience that it is only now that Gallileo is being recognized by some people.

Foster: Perhaps I can add something. We have also been busy with such resistance effects. And I think some members of our group would support you. We will have to think this over.

176

Heart-Cycle-Synchronized Jet Ventilation

H. T. van der Zee, N. S. Faithfull, and W. Erdmann

Conventional mechanical ventilation (CMV) can readily produce haemodynamic changes in both the pulmonary and systemic circulations, and obvious interaction can be observed between intratracheal and intra-alveolar pressures and the circulatory state of the patient. High-frequency ventilation has often been claimed to result in gas exchange of equal efficiency to, if not better than, that which can be achieved with CMV. At the same time, peak and mean intratracheal pressures are lower than those observed during CMV. The main factor in determining the degree of interaction between circulation and mechanical ventilation is the level to which the mean tracheal pressure rises. Numerous studies have demonstrated that increasing positive end-expiratory pressure (PEEP) may have deleterious effects on circulatory performance. Few studies have been performed examining the moment of insufflation in relation to the phase of the heart cycle and the consequences that this may have for the pulmonary and systemic circulations.

Since the advent of high-frequency ventilation in clinical practice, it has become clear that the most widely used frequencies of ventilation are in the range of 2 Hz. It is thus clear that the frequency of jet ventilation applied to a particular patient may fall well within the range of the heart rate. A study of jet ventilation coupled to heart cycle was therefore carried out to determine whether a predictable interaction could be found between the ventilatory pattern and the state of the pulmonary and systemic circulations.

Methods

The experiments were performed in sheep which had been instrumented the day before the actual experiment. Pressures were measured in the internal mammary artery (AP), the pulmonary artery (PAP) and the left atrium (LAP). Aortic flow (AF) and pulmonary flow (PF) were monitored using electromagnetic flow probes. The electrocardiogram (ECG) was continuously measured and intratracheal pressure (TP) was monitored.

Anaesthesia was induced with thiopentone, and after intubation the animals were paralysed and ventilated with N_2O and O_2, doses of fentanyl being administered when necessary. The jet ventilator used has been described in the chapter by T. Mostert and W. den Dunnen. Briefly, this jet ventilator has the ability to deliver insufflation synchronous with or at various fixed time intervals from the ECG-QRS complex. Inspiration/expiration ratios can be varied over a wide range.

Initially, conventional ventilation was established in order to measure normal control values of haemodynamics and gas exchange. The experiments were commenced using intermittent positive-pressure ventilation (IPPV) while maintaining normal arterial $PaCO_2$ and PaO_2 levels.

After control measurements had been performed, jet ventilation was commenced using a cannula inserted into the endotracheal tube and placed about 10 cm above the carina. A rate was chosen closely approximate to the heart rate, but asynchronous with the ECG. The jet ventilator switched to a mode synchronous with the ECG, using two distinct patterns. With the first setting insufflation started after the ejection of blood into the pulmonary circulation had taken place. Insufflation time thus coincided with ventricular diastole or lung capillary blood transit period. This is termed "counterphase jet ventilation". With the second setting insufflation was commenced immediately prior to the ejection of blood into the pulmonary circulation, and hence insufflation coincided with ventricular systole. This is termed "in-phase jet ventilation". Peak intratracheal pressures were held below that of CMV, and tidal volumes were adjusted to achieve normal arterial PaO_2 and $PaCO_2$ levels. The haemodynamic parameters were continuously monitored. The inspiration/expiration ratio was $1:1$.

Results

During IPPV the familiar interference pattern between peak inspiratory pressures and systemic and pulmonary haemodynamics could be observed and an oscillatory pattern could be seen in all the monitored haemodynamic parameters, i.e. systemic and pulmonary arterial pressures, arterial pressure, PF and AF (Fig. 1). During inspiration there was an increase in LAP and PAP. At the same time, one could observe a decrease in PF; an increase in AF was followed by a decrease. With increasing end-respiratory pressure, there was a further increase in PAP but not in LAP. The PF continued to decrease and was almost reduced to zero during inflation when end-expiratory pressure reached 12 mm Hg. At the same time, AF progressively decreased (Fig. 2). The LAP showed only a small increase with increasing PEEP.

When jet ventilation asynchronous with the ECG but at a rate approximately equalling the heart rate was employed, an oscillatory pattern was again seen, but in a more condensed form (Fig. 3). Analysis of this haemodynamic interference pattern reveals a first harmonic oscillation determined by the difference in frequency between the heart rate and the jet frequency. It should be noted that peak insufflation pressures were considerably less than those seen during conventional ventilation, while the same efficiency of gas exchange was maintained. During this type of jet ventilation, the lowest systolic pulmonary pressures and highest pulmonary diastolic pressures and the lowest amplitude in LAP wave were found when insufflation pressure rise corresponded with injection of blood into the pulmonary capillary bed. A slight decrease in pulmonary flow was also seen at this time. The changes in AF were opposite to those in PF.

One hundred and eighty degree phase-shifting of the ECG-triggered ventilation (in-phase jet ventilation), gives the opposite results: increase in peak pulmonary systolic pressure and lower pulmonary end-diastolic pressures; increase in LAP wave amplitude; and increase in PF and decrease in AF. A clearer picture of the haemodynamic events could be seen when jet insufflation occurred with ev-

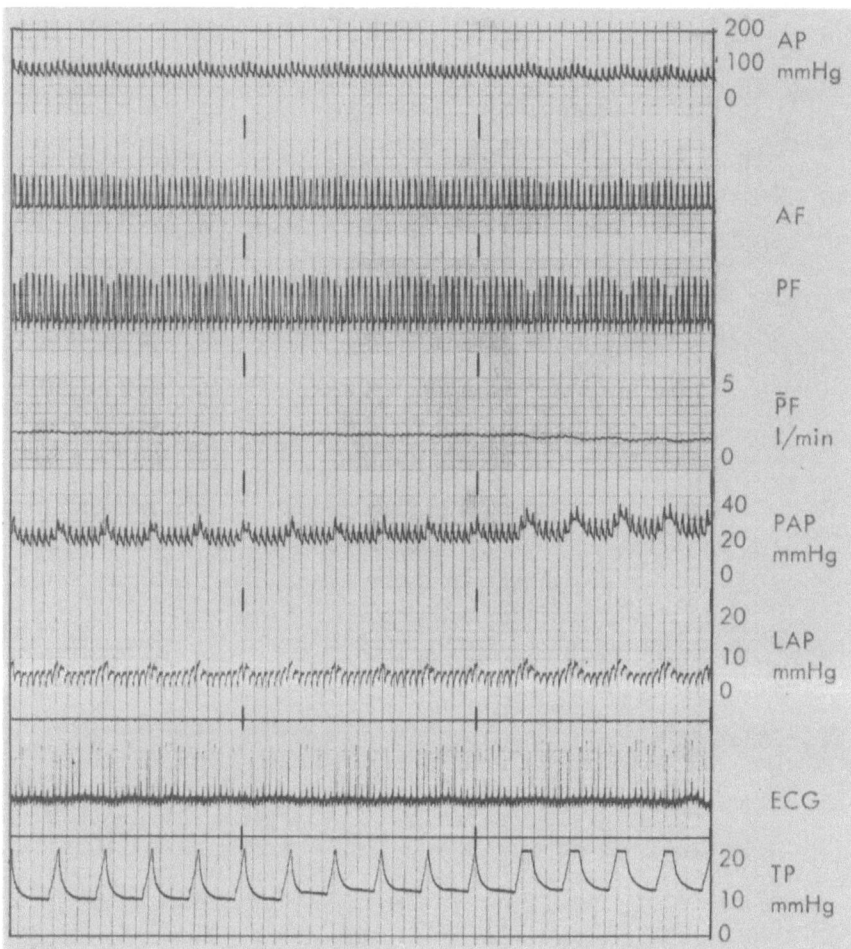

Fig. 1. Tracings with the interaction between intermittent positive-pressure ventilation (IPPV) and the haemodynamic state during progressively increased positive end-expiratory pressure (PEEP). Shown are internal mammary artery pressure (*AP*), aortic blood flow (*AF*), pulmonary artery blood flow (*PF*), pulmonary arterial pressure (*PAP*), left atrial pressure (*LAP*), electrocardiogram (*ECG*) and intratracheal pressure (*TP*)

ery other beat instead of every heart beat (Fig. 4). Here counterphase insufflation produced the same interference pattern: a decrease in PF; an increase in AF; a decrease in peak systolic PAP and an increase in end-diastolic PAP; and an increase in LAP wave amplitude.

If this counterphase ventilation synchronous with the ECG took place with every single heart beat, then the following haemodynamic phenomena were repeatedly observed (Fig. 5). As soon as the jet ventilation was initiated in the counterphase setting after conventional ventilation was stopped, there was an immediate increase in both PF and AF (an increase of 60%–70%). There was also an immediate rise in LAP and PAP. These initial increases in LAP and PAP took place over a very short time (not more than a few minutes). This increase in flows and pressures was not maintained, and after the initial increases the pressures progres-

179

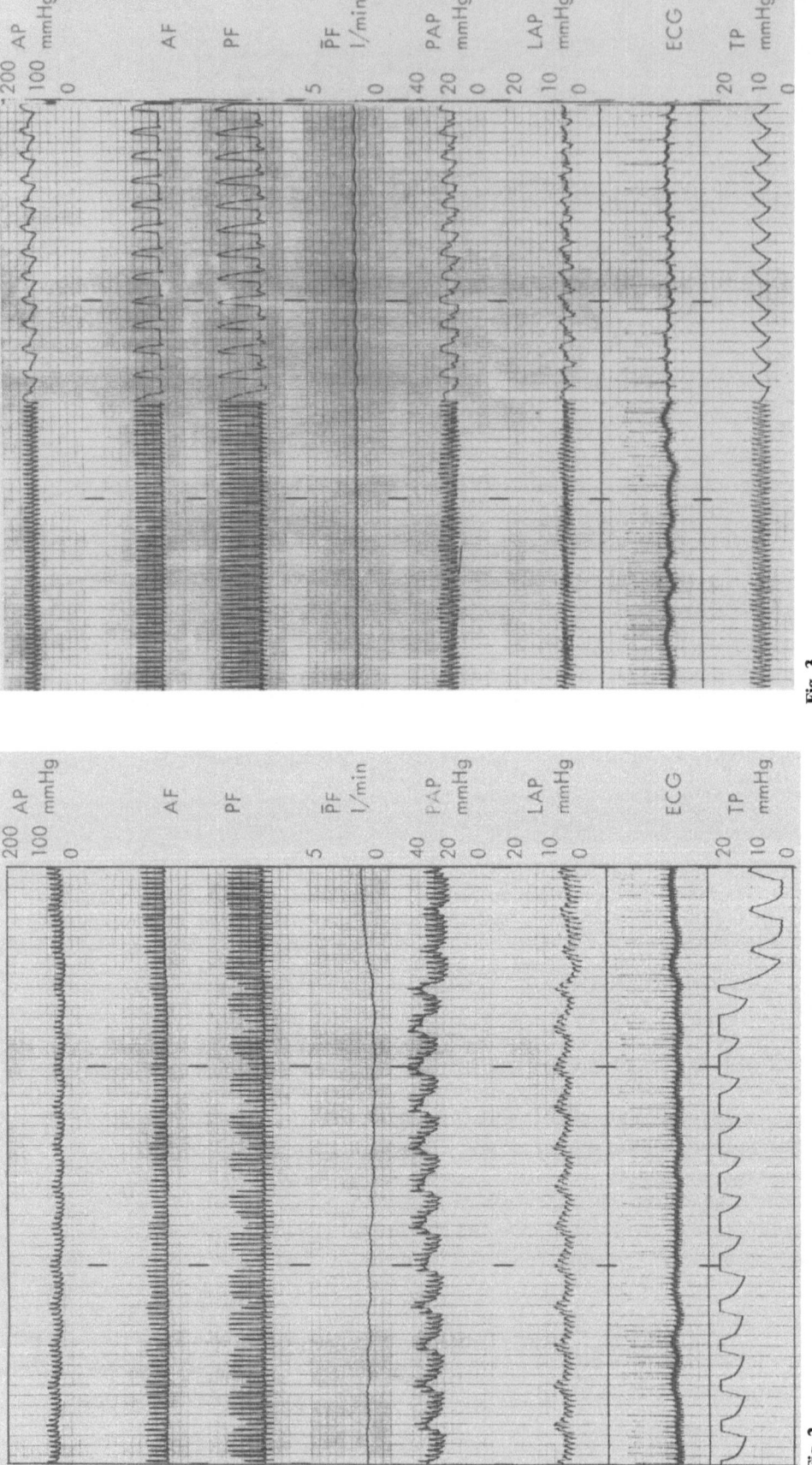

Fig. 2 The same tracings as in Fig. 1. PEEP has been increased to 12 mmHg. Note the severe reduction in PF and increase in PAP during inflation

Fig. 3 Haemodynamic interference pattern during jet ventilation at a rate equal to the heart-rate but asynchronous with the ECG

Fig. 3

Fig. 2

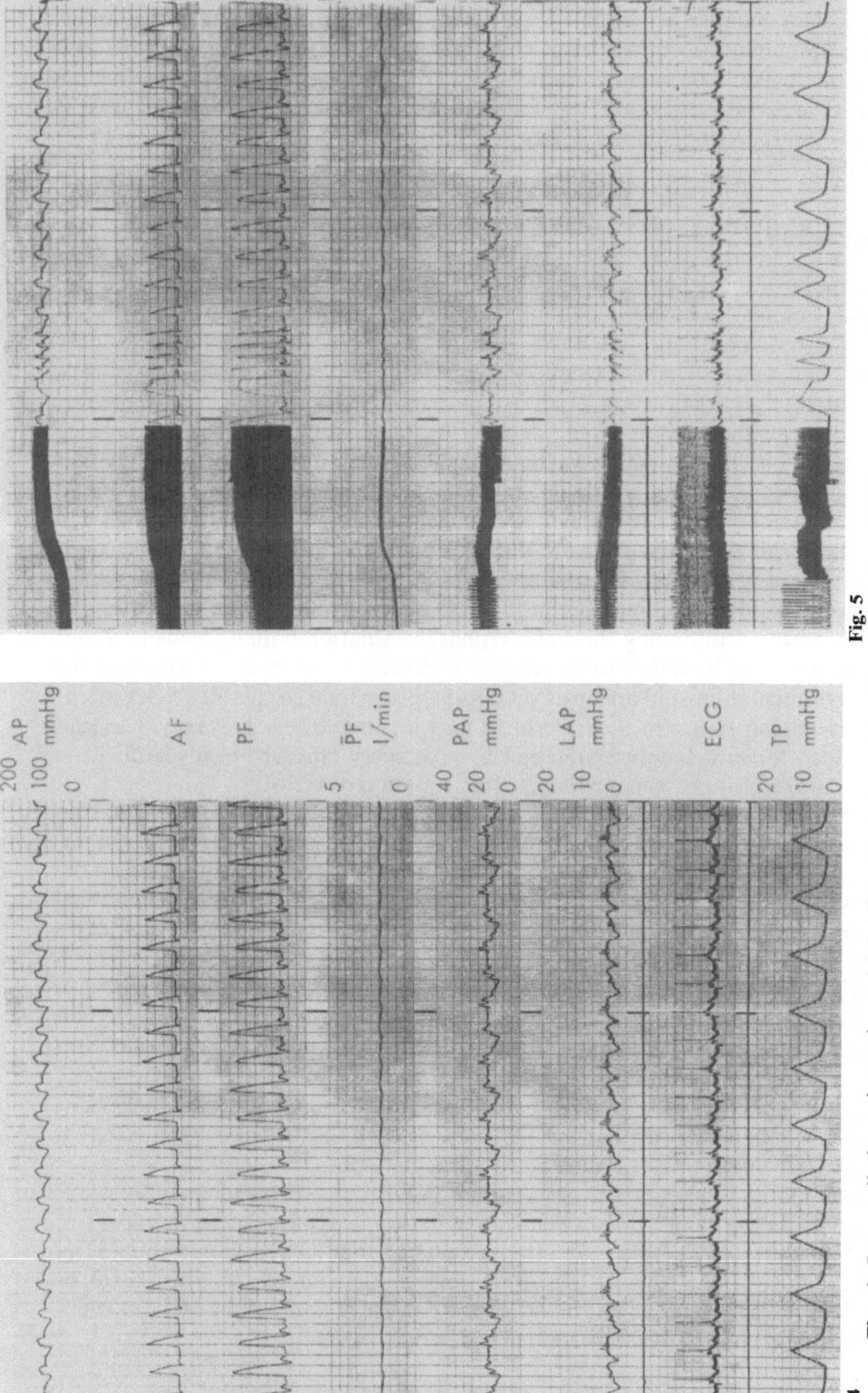

Fig. 4. Jet ventilation triggered on the ECG, but at every other beat. The rise of TP occurs after the peak PE, measured as the root of the pulmonary artery, but coinciding with the lung capillary blood transit period. (Counterphase jet ventilation.)

Fig. 5. Counterphase jet ventilation synchronous with the ECG but at every beat. Note the immediate increases in AP, AF, PF, PAP and LAP. PAP and LAP decline after the initial increase and severe bradycardia develops

sively fell to a point where the LAP fell well below the value obtained during CMV. If this counterphased ventilation was allowed to continue, the induced haemodynamic changes often resulted in severe bradycardia, and occasionally in cardiovascular collapse. A possible explanation for this phenomenon is that an increased pumping effect takes place with this counterphased setting of ventilation. Thus blood would be pumped from the right heart to the left heart, increasing the LAP. The initial rise in LAP was followed by a decrease in LAP due to lack of venous return. Arterial blood gases taken during these runs showed normal PaO_2 and $PaCO_2$ values.

Discussion

The results of the experiments show that high-frequency jet ventilation produces definite changes in both the pulmonary and the systemic circulations. These effects are both short and long term. A clear interference pattern between pneumatics and haemodynamics can be seen when the jet ventilation rate closely approximates to that of the heart rate. An undulating pattern of haemodynamic alterations can then be observed, showing characteristics of the beat frequency (harmonic oscillation) which are determined by the difference between ventilatory rate and heart rate. These pneumatically induced maximum and minimum haemodynamic alterations can be augmented by programming the jet insufflation to a particular fixed time frame of the QRS complex. In particular, at the counterphase setting of ventilation (rising alveolar gas pressure coinciding with pulmonary capillary blood filling), a considerable haemodynamic interference can be demonstrated. Continuing this counterphase ventilation initially leads to a rapidly augmented blood flow to the left heart. However, due to the hysteresis of the systemic circulation, and due to as yet poorly understood circulatory reflex mechanisms, venous return to the right heart may be diminished. This may then lead to sudden forward failure manifested by repeatedly observed bradycardia. The bradycardia disappears immediately after initiation of IPPV.

Another possible consequence of this haemodynamic event may be a sudden increase in cardiac output and, consequently, transmission of the large pulmonary pressures to the pulmonary capillaries. This may then cause damage to the endothelium of the lung capillaries. Autopsy in one animal in the series revealed haemorrhagic oedema of the lung following prolonged counterphased ventilation which led to increased pulmonary pressures.

In conclusion, our results indicate the possibility that counterphased jet ventilation may have a considerable influence on the pulmonary and systemic circulations. It is also conceivable that certain cardiac patients might benefit from counterphased jet ventilation if care is taken that the time of insufflation is properly triggered to the heart cycle. Such patients might be those with right ventricular failure or pulmonary valvular insufficiency, or even those whose systemic circulation was being supported by an intra-aortic balloon.

It should be noted that this jet ventilation technique presents a considerable risk in unskilled hands, even when jet ventilation is applied using very low tracheal pressures. We would therefore recommend that jet ventilation should not be used on patients with a jet rate approximating the patient's own heart rate, or first harmonic thereof, as this increases the danger of interaction between ventilation and the haemodynamic state of the patient.

High-Frequency Ventilation for Laser Surgery of the Larynx

P. A. Scheck, C. Mallios, and P. Knegt

Surgery of the larynx via the peroral endoscopic route using the laser beam has introduced new possibilities in the treatment of different pathological processes in this area. In comparison to other surgical methods, endolaryngeal surgery using the laser beam has several advantages but also involves a serious risk.

The advantages are:

1. There is minimal bleeding because blood-vessels of less than 0.5 mm in diameter are coagulated. Even oedema is minimal, as cells near the impact point are not affected. The consequence of these characteristics is that tracheostomy may frequently be prevented in cases where it would be necessary.
2. Healing is nearly painless and is precise and rapid because there is little damaged tissue and epithelial migration proceeds rapidly.
3. Functional results are very good, as there is minimal contraction or scar tissue formation.

The main risk in the use of the laser beam is the ignition of the endotracheal tube with the risk of burns and the inhalation of smoke (Schramm). Therefore, during laser surgery of the larynx the anaesthetist has to deal with special circumstances and conditions in ensuring a free airway and adequate ventilation of the fully paralysed patient.

A standard armoured endotracheal tube with an internal diameter of 7 mm or more fills the glottis to such an extent that it is difficult for the surgeon to find sufficient space to carry out the treatment properly. There are, of course, no problems when the patient is ventilated with a tracheostomy cannula.

The laser beam can cause ignition of practically all cannulas and tubes used in the upper airway for ventilation or for maintaining a free airway. In order to demonstrate the effect of the laser beam on various materials we have irradiated some of the materials commonly used in the upper airway. Endotracheal tubes react either by bursting in flames (in the case of armoured tubes) or by melting of the outer layers, which is what happens to red rubber tubes. Portex tubes produce dense smoke. Vyggo endotracheal tubes react in a similar way. The 14-gauge catheter which we use for high-frequency ventilation produces dense smoke as well. An aluminium foil covering the same catheter on the focusing point of the laser apparatus completely protects against ignition by the laser beam even when this beam hits the foil several times in the same place. The catheter has to be covered by the foil as close as possible to its tip, leaving at least two side-holes of the catheter free.

Table 1. Bloodgases (kPa)

	60 min^{-1} ($n=112$)	90 min^{-1} ($n=23$)
pH	7.32±0.15	7.37±0.16
PaO$_2$	16.3 ±1.2	16.6 ±1.5
PaCO$_2$	4.1 ±0.9	4.5 ±0.6
Sat. O$_2$	98.3 ±0.5	97.7 ±0.9

A flame in the patient's airway is extremely dangerous and may even endanger his life. We had experience with more than 100 high-frequency ventilations carried out using the Bronchovent apparatus with blood gas monitoring before we used this technique for laser surgery of the larynx. Anaesthesia for these procedures is described elsewhere. The technique of ventilation is as follows: The 14-gauge insufflation catheter is, as a rule, passed through the nostril and with the aid of a Magill forceps directed into the trachea. The catheter is then connected to the ventilator. The Kleinsasser instrument is fixed and surgery on the larynx can commence. To prevent damage to the nasal mucosa by the aluminium foil, the nasal catheter may be introduced through a nasal airway.

The insufflation catheter leaves almost the whole glottis free and is even out of the surgeon's view. A disadvantage of the use of this small catheter during high-frequency ventilation is the impossibility of protecting the sublaryngeal region or trachea from the laser beam with moist cotton or gauze. This would, of course, impede expiration. If necessary, the laryngeal structures can be protected by a metal mirror.

Another technique which might be useful under difficult anatomical conditions is to ventilate through a metal catheter with an internal diameter of 2.5 mm or more. Air entrainment occurs during ventilation through this metal catheter, which is without side-holes. Carbon dioxide values increase slowly and ventilation should not exceed 30 min. In some patients, it is even technically possible to measure end-expiratory CO$_2$.

High-frequency ventilation together with laser treatment has been used so far for 65 treatments in the larynx, for Zenker's diverticalum of the oesophagus in ten patients and for the laser treatment of tumours of the base of the tongue in six patients.

The blood gas values are summarized in Table 1. When a mixture of air with 50% oxygen is used for ventilation, oxygen saturation remains unchanged. On the other hand, as expected, during ventilation with a small tidal volume, carbon dioxide tension increases slightly after 1 h in some patients. For evacuation of the smoke during laser treatments, a second catheter is placed nasally with its tip in the nasopharynx. A low suction pressure is effective in producing a practically smoke-free surgical field.

Discussion XV

Foster: Dr. Scheck, do you use PVC catheters? PVC makes a very poisonous gas when it is burned.

Scheck: We have two sorts of catheter. One is made out of metal and is used very close to the area where the laser is working. Additionally, we have a second catheter in the nostril, which exhausts all the burnt gases.

Foster: I was just wondering if you were using a new plastic which is safer than PVC? We would be very happy to have one.

Scheck: We are in contact with Dr. Wainright from Southhampton who is developing something new. Recently, something new was developed in the USA, but it is impossible to get it here.

McIntyre: Did you ever have any problems with an undesirable increase in intra-pulmonary pressure?

Scheck: Yes, especially if the catheter is inserted too deeply.

McIntyre: This is also our experience. Each one should have appropriate safety devices.

The Pre-, Intra-, and Postoperative Organization of Anaesthesia [1]

R. Droh

Informing the Patient

Right at the beginning of a clinical stay we can give the patient good and simple information with a patient information booklet (*Patienten-Aufklärungsbogen*, Fig. 1). This booklet helps to avoid many repeated questions and answers, waste of time, fears and disorganization, and we have more time to answer individual questions. So it creates a good relationship between doctor and patient right from the beginning.

We have developed this information booklet within a period of 10 years. During this time we have used it on about 65 000 patients with good results and with very good feedback from the patients. It shows how one can establish a systematic means of informing the patient. Of course, every department must have its own special information booklet.

It informs the patients about the location, the doctors, the paramedical personnel, the function, the methods, the security, the organization, and the most important rules and provides other information about the anaesthesia department wanted by the patient.

Anamnesis

Also over a period of 10 years we have been taking the patient's anamnesis with the help of an anamnesis questionnaire (*„Krankenfragebogen"*, Fig. 2). Parents can answer this questionnaire for their children. Foreigners or handicapped patients are aided by our nurses to answer the questionnaire, as are the parents if they have problems with it.

In the daily routine work of our anaesthesiological pre-operative visit, we always have to ask the same questions and we get tired of continuously repeating them. At the same time we often have only a short time for the pre-operative visit. It was therefore a logical step to create a standardized questionnaire for anamnesis. Of course, the scheme of question and answer in such a questionnaire must be established in such a way that one can immediately recognize positive answers. One can thus save time to handle particularly interesting points more intensively.

---

[1] Figures listed on pp. 190–217. Figure 1 with English summary on p. 189. Figure 7 translated from the German

Frequent changes of the nurses, shift-service and also a certain superficiality on the part of the paramedical staff make it additionally necessary to give the patient written information. In urgent cases too the patient must receive basic information. We therefore also added a short informational passage for the patient as an introduction to our anamnesis questionnaire. The questionnaire closes with the patient's declaration that he agrees to the suggested anaesthesia and his statement that he has answered truthfully. The whole is signed by the patient.

Document Recording Laboratory Values and Results of Other Medicotechnical Investigations

Besides anamnesis and clinical examination we routinely need a certain number of laboratory values. We have developed a standard laboratory document (Fig. 3) which at the same time serves as a standardized medical report, as well as providing information within the clinic: it sums up all medicotechnical parameters. With it we can transfer the collection of necessary information (laboratory, X-ray, clinical examination etc.) to our paramedical team; our work is thus more rationally organized. The laboratory document gives us a clear synopsis and enables us very quickly to find missing or pathological parameters.

A further advantage of this document is that it speeds up communication between the different departments (with the laboratory, the X-ray department etc.). At the same time duplication of examinations is avoided, thus lowering costs.

This document is designed in such a way that it supports us in selecting the necessary examinations. It covers not only anaesthesia but also intensive care and it will also be used as a source of reference for colleagues inside and outside the hospital.

The Pre-operative Visit and Outpatient Clinic

In order to maintain professional secrecy we should not question a patient in the presence of other patients. Every patient has the right to the absolute security of our professional secrecy when he reveals his most private problems to us. It is necessary to have a separate room for the pre-operative visit. Connected with this room should be a separate secretariat so that the security of written information about every patient is also guaranteed.

Every patient should receive his documents in a closed portfolio. He takes his portfolio with him back to his room or to other departments. However, one should not make too much of a mystery about his papers, but give him the chance to have a look at his documents. We use a portfolio which the patient receives when he enters the clinic. This portfolio accompanies the patient to the pre-operative visit, to his ward, to the operating theatre and to the intensive care unit.

Before the pre-operative visit is begun, our paramedical team has to obtain the following:

1. The patient information booklet *(Patienten-Aufklärungsbogen)*, signed by the patient
2. The anamnesis form *(Krankenfragebogen)*, signed by the patient

3. The document containing the laboratory values and results of other medico-technical investigations, such as X-rays and ECG
4. The anaesthesia protocol

All these requisite data are checked by a nurse. If they are not complete and correct they must be completed by the nurse with the help of the patient and/or his relatives. With the medical document we obtain the following basic laboratory values: haemoglobin, haematocrit, erythrocytes, blood coagulation, urine status and urine sediment, a chest X-ray and an ECG. When we start the pre-operative visit the patient is already lying on an examination table. The papers are put in order and the X-rays are placed in the light-box.

If additional investigations are necessary, they are ordered via the medical document, which accompanies the patient to the laboratory and other places. Communication between the different departments is thus very well guaranteed without further extensive inquiries.

The Anaesthesia Protocol

The anaesthesia protocol (Fig. 4) contains pre- and intra-operative findings, standard instructions and details of the most frequently used drugs. Important clinical and anamnestic peculiarities are transferred from the medical document to the protocol.

The Selected Methods of Anaesthesia

The selected methods of anaesthesia will be transferred to the anaesthesia part of the operating programme. This is the basis (Fig. 5) on which the paramedical team prepares the different groups of patients a while before the normal operating programme is begun; it can be done in the afternoon of the day before or about an hour before anaesthesia and surgery are begun.

Control of the Preparation for Anaesthesia

Shortly after the patient arrives in the operating theatre we use a check-list (Fig. 6) to check up the patient for his anaesthesia and operation. The nurses must always stick to the instructions on the check-list. In this way no important question will be disregarded. The nurse has to confirm the remarks on the check-list with her signature. The check-list remains in the portfolio of the patient and has to be registered together with it.

Check-list for Control of Anaesthesia and Monitoring Equipment

Another check-list (Fig. 7) guarantees that every monitor and all the rest of the equipment is ready for use before anaesthesia starts. At the same time it is a standardized guide-line for the work of our nurses. It is a great improvement for our security and the security of our patients.

The Recovery Room

In the recovery room we take care of our patients after anaesthesia. It is one of the most important check-points for our security and the security of our patients. There is a certain limit for the personal, technical and medical equipment of this room which must be guaranteed. Every task must be performed following certain instructions, which are prescribed in a special check-list (Fig. 8). We recommend that a recovery room book *(Aufwachraumbuch)* also be kept. This book not only serves as a standard instruction for the work in the recovery room, but is also a document of its execution and results.

The Postanaesthesia Questionnaire

Twenty-four hours after recovery we ask every patient about his experiences with the department, premedication, recovery, pain relief, special wishes etc. by means of a questionnaire (Fig. 9).

References

Droh R (1975) Ein Patientenfragebogen zur Rationalisierung der anaesthesiologischen Befunderhebung. Anaesth Praxis 10:25–27
Droh R (1977) Der Aufwachraum, Organisation und Ausstattung. Anaesth Praxis 13: 143–147
Droh R (1977/1978) Regionalanaesthesien. Anaesth Praxis 14:10–12
Droh R, Rothmann G (1979) Anaesthesiologisch wichtige Laborwerte und Untersuchungsbefunde in einem standardisierten Arztbrief. Anaesthesiol Intensivmed 20: 26–30
Heiderhoff K, Droh R (1977) Organisation der kontinuierlichen Periduralanaesthesie. Anaesth Praxis 13:35–47

Fig. 1 (Summary). The information booklet ("Patienten-Aufklärungsbogen") provides information about the following points:

1. It contains a map of the clinic which enables the patient to find the different rooms (laboratory, X-ray, ECG) very easily.

2. It gives guide-lines for the pre- and post-operative period.

3. It provides a descriptive explanation of the course of narcosis.

4. It contains some information about the general risk of narcosis.

5. It explains several points which often give reason for complaints or questions (side-effects of narcosis, preoperative waiting-time, etc.).

6. It proves information and guide-lines for out-patients, too.

7. The patient has to reconfirm the contents of the information-booklet by signing it.

A N A E S T H E S I O L O G I E

Sportkrankenhaus Hellersen
Chefarzt : Dr. R. Droh

Adressette (bzw. Name [bei Frauen auch Mädchenname], Vorname, Geb.-Datum, Adresse)

Patienten - Aufklärungsbogen

Sehr verehrte Patientin
Sehr geehrter Patient

Dieses Informationsblatt soll Sie über die Narkose im speziellen und die Arbeit der Anaesthesiologie des Hauses im allgemeinen aufklären.

Der untenstehende Plan ist als Orientierungshilfe für Sie zum leichteren Auffinden unseres Arbeitsbereiches gedacht.

Am Ende des Flures, der vom Haupteingang ausgeht, finden Sie einen **Warteraum** und einen **Untersuchungsraum** oder auch **Vorstellraum** genannt, unter der Bezeichnung **„Anaesthesie"** das Chefsekretariat sowie Untersuchungs- und Arzträume. Direkt gegenüber der Auskunft am Haupteingang liegt das **„Labor"** und neben dem Aufzug, vor dem Operationsbereich, der **„EKG"-Raum.** Dieser Bereich einschließlich der **Intensivmedizin** umfaßt den Arbeitsbereich der Anaesthesiologie.

Eine der **Hauptaufgaben** der Anaesthesie ist die **Schmerzausschaltung** für operative Eingriffe. Diese Schmerzausschaltung wird entweder mit einer **Allgemein-Anaesthesie** auch **Narkose** genannt, oder einer **Lokal-Anaesthesie,** auch **Örtliche-Betäubung** genannt, erreicht. Hierzu werden eine Reihe von Medikamenten verwandt, die am Gehirn und den Nerven des Menschen ihre Wirkung entfalten.

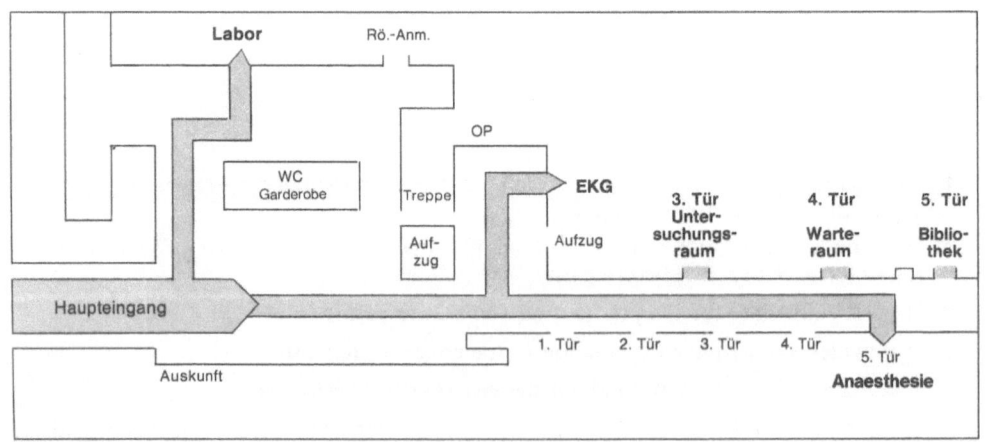

Fig. 1   *(will be continued)*

190

Eine andere Aufgabe ist z. B. die **Behandlung von Herz und Kreislauf** unter besonderer Einbeziehung von so lebenswichtigen Organen wie **Leber, Nieren und Drüsenorgane** (z. B. Bauchspeicheldrüse, Schilddrüse, und Nebennieren) die ja ebenfalls von den Narkosemitteln erreicht werden.

Mit der modernen Narkose werden heute einerseits die verschiedenen Funktionen des Körpers entsprechend den operativen Bedürfnissen gesteuert bzw. ausgeschaltet und alle vitalen Funktionen des Körpers zum Schutz des Patienten künstlich geregelt übernommen. Die Narkose wird außerdem dem Gesundheitszustand des Patienten und seiner persönlichen Konstitution angepaßt.

Zur Erfüllung dieser Aufgaben ist ein erheblicher Aufwand von Ärzten und anderem medizinischen Personal sowie Technik erforderlich. Diese Aufgabe wird von uns unter dem höchsten Sicherheitsprinzip erfüllt. Für uns gibt es keine großen oder kleinen Narkosen, sondern eben nur Narkosen. Dies bedeutet, daß die Größe des operativen Eingriffs für unsere Arbeit nur von sekundärer Bedeutung ist.

Der Ablauf einer Narkose ist, allgemeinverständlich ausgedrückt, mit dem Fliegen vergleichbar. Auch hier wird an die Sicherheit ein sehr strenger Maßstab angelegt, wobei die geforderte Flugsicherheit zunächst nichts mit der Flugstrecke zu tun hat. Eine Flugstrecke über Meere, Wüsten, Eisregionen, Städte und Berge wäre einer Narkose zu verschiedenen operativen Eingriffen vergleichbar. Das Landen und Starten beim Fliegen entspricht sehr gut dem Ein- und Ausleiten der Narkose. Fliegen und Anaesthesie sind auch für den technischen Stand der Geräte und dem menschlichen Können vergleichbar.

Darum entspricht auch der Sicherheitsaufwand für die Anaesthesie durchaus dem der Fliegerei, wobei nur ein sehr geringer Teil davon sichtbar wird, während der größte Teil für die Patienten ebenso wie für die Passagiere verborgen bleibt und beide Verfahren erfordern viel Einsichtigkeit, Geduld und volles Vertrauen.

Wir bitten Sie auch darum um Verständnis, wenn wir Ihnen im Rahmen unserer Aufklärung keine Einzelheiten anzubieten versuchen, denn so wenig wie ein Flugkapitän seinen Passagieren das Cockpit, das Fliegen oder seine Entscheidungen erklären kann, so wenig ist uns dies für die Patienten möglich. Sie kennen das Ziel, die für Sie vorgesehene Operation. Die Narkose ist das dafür vorgesehene „Flugzeug", wofür eine Vielzahl von Geräten erforderlich ist, die vom speziell geschulten Personal gewartet und bedient werden müssen. Ein Teil dieser Geräte ist zur Verabreichung der Betäubung notwendig, ein anderer Teil zur ständigen Sicherheitsüberwachung wichtiger Körperfunktionen und wieder ein anderer Teil zur Unterhaltung der lebenswichtigen Funktionen des menschlichen Körpers

Die Verantwortung für diese Arbeit wird in der Fliegerei von dem Flugkapitän und bei uns von dem Chefarzt getragen. So wie es dort Copiloten, Techniker und Bordpersonal gibt, haben wir Oberärzte, Med. tech. Assistentinnen, Pfleger und Schwestern.

Die Ärzte für Anaesthesie in unserer Klinik sind:

Die der Operation und Anaesthesie vorausgehenden Untersuchungen und Befragungen, auch über Krankenfragebogen, erbringen die für die Auswahl des Betäubungsverfahrens notwendigen Informationen. Gelegentlich muß auf Grund der hier gewonnenen Ergebnisse eine Operation oder Anaesthesie zurückgestellt werden, um durch weitere Untersuchungen noch notwendige Informationen zu beschaffen. Damit wird die Sicherheit, Gesunderhaltung und Gesundung des Patienten gewährleistet, gelegentlich kann statt einer Vollnarkose ein örtliches Betäubungsverfahren empfohlen werden.

Sollten wir Ihnen ein örtliches Betäubungsverfahren vorschlagen, so ist vorab schon dazu zu sagen, daß auch die örtlichen Betäubungsverfahren natürlich ebenso wie die Allgemeinanaesthesieverfahren (Vollnarkose) ihre Vor- und Nachteile sowie Risiken haben.

Nachteile des örtlichen Betäubungsverfahren sind z. B.:

a) daß Sie während der Operation bei Bewußtsein sind und sich damit der Situation seelisch gewachsen fühlen müssen;

b) die Vorbereitungen in Ihrem Bei- und Wachsein länger dauern als bei der Allgemeinanaesthesie;

c) es dabei ganz wesentlich auf Ihre Mitarbeit ankommt;

d) auf Grund von anatomischen Schwierigkeiten die Lokalanesthesieverfahren gelegentlich versagen können und dann durch ein Allgemeinanaesthesieverfahren ersetzt werden müssen;

e) im Falle des Versagens die Operation aber um einen Tag verschoben werden kann;

Fig. 1 (contd.)   *(will be continued)*

f) bei örtlichen Betäubungen der volle Wirkungseintritt je nach Verfahren 15-30 Minuten dauern kann; (Selbstverständlich prüfen wir vor Operationsbeginn in jedem Fall, daß Sie auch absolut schmerzfrei sind).

g) die Schmerzfreiheit bei örtlichen Betäubungen gleichzeitig zu einer ebenso lang andauernden Lähmung des örtlich betäubten Bereichs kommt.

Von Vorteil ist bei den örtlichen Anaesthesien unter anderem, daß die Schmerzfreiheit 3 - 4 Stunden und länger anhält.

Sollten wir uns einmal ausnahmsweise für ein örtliches Betäubungsverfahren entschließen, dann wird dies mit Ihnen während der anaesthesiologischen Voruntersuchung **vor der Operation** ausführlich besprochen und Ihre Einwilligung hierzu eingeholt.

Am Abend vor jeder Anaesthesie bekommen Sie ein Medikament zur Narkosevorbereitung, das Sie körperlich und seelisch entspannt. Am Morgen vor Narkosebeginn wird Ihnen auf der Station eine Spritze zur weiteren Narkosevorbereitung verabreicht.

Diese beiden Medikamente bilden zwar die Grundlage für die vorgesehene Anaesthesie, dennoch dürfen Sie davon keine betäubende Wirkung erwarten. Denn wir wollen Sie so schonend wie nur irgend möglich mit Medikamenten behandeln.

Nach Erhalt Ihrer Vorbereitungsspritze werden Sie nach Ablauf etwa einer halben Stunde von Ihrer Station in den Vorraum zum eigentlichen Operationstrakt gebracht. Von dort geht es über eine hydraulische Schleuse dann in den Narkoseeinleitungsraum. Hier werden nun alle Vorbereitungen für die Einleitung der Narkose getroffen: Aufzeichnung von Pulskurve und EKG, Blutdruckmessung und Anlegen einer Infusion. Durch diese Infusion erhalten Sie alle weiteren Medikamente, so daß wir Sie nur einmal zu „pieken" brauchen.

Sind diese Vorbereitungen abgeschlossen, leitet der Anaesthesist die Narkose mit einer „Schlafspritze" ein. Bevor Sie einschlafen, beginnt Ihr Herz etwas schneller zu schlagen, Sie sehen vielleicht Doppelbilder und Ihre Augen und der Mund können ein wenig trocken werden – alles Reaktionen, die wir absichtlich zur Vervollkommnung Ihrer Sicherheit herbeiführen. –

Erst wenn Sie eingeschlafen sind und die Narkose voll wirkt, werden Sie in den Operationssaal gebracht.

Nach der Operation können gelegentlich durch die Trockenheit in Mund und Hals Durstgefühle und Schluckbeschwerden auftreten. Wie nach einem ganz normalen Schlaf ruft die verminderte Speichelabsonderung einen faden Geschmack hervor; aber auch dies ist völlig normal und kein Grund zur Beunruhigung.

Nach Abklingen der Narkose werden im Aufwachraum häufig schmerzstillende Mittel verabreicht, die Sie noch für Stunden weiterschlafen lassen. Falls Sie nach der Narkose aber unbedingt hellwach sein wollen, so sagen Sie es uns bitte. Wir werden Ihren Wunsch respektieren. Sie müssen dann allerdings auch den Wundschmerz ertragen, da das eine das andere aufgrund der „Konstruktion" des Menschen ausschließt.

Nur wer leicht unter Übelkeit leidet, sollte nach der Narkose möglichst wenig essen und trinken, zumal die Durstgefühle durch Trinken nicht völlig beseitigt werden können und Leber, Darm sowie die übrigen inneren Organe selbst unter normalen (physiologischen) Schlafbedingungen Ihre Arbeit auf ein Mindestmaß beschränken, was eben auch für den Schlaf nach der Allgemeinnarkose zutrifft.

Zur Operations- und Anaesthesie-Vorbereitung sowie zur Behandlung danach haben wir für Sie einen Ernährungsplan erarbeitet, der die neuesten wissenschaftlichen Erkenntnisse der Klinik ebenso berücksichtigt wie die der Weltraumforschung. Diese Spezialnahrung oder auch „Astronautennahrung" genannt enthält die notwendigen Eiweißmengen, Spurenelemente (Zink, Magnesium, usw.) Vitamine, Elektrolyte, Aufbaustoffe und alle anderen notwendigen Nahrungsanteile (Fette und Kohlenhydrate) in der Zusammensetzung und Menge, wie sie der Körper zum Leben, Wohlbefinden, zur ungestörten, raschen Wundheilung sowie Gesundung braucht und am einfachsten aufnehmen sowie zur Verfügung stellen kann.

Fig. 1 (contd.) *(will be continued)*

Diese „Astronautennahrung" wird heute so perfekt hergestellt, wie dies keine herkömmliche Küche zu tun vermag. Als Getränke stehen „Mineraldrinks" mit den notwendigen Salzen zur Verfügung. Selbstverständlich ist eine solche Kost nicht zur Dauerernährung eines gesunden und völlig intakten Menschen gedacht und notwendig, aber für einen durch Anaesthesie und Operation beeinflußten Körper ist sie eine echte Entlastung, die die Funktionstüchtigkeit des Körpers verbessert.

Die Nahrung ist aber nicht nur aus den richtigen Bausteinen und Elementen optimal zusammengesetzt, sondern auch noch schlackenarm und leicht verdaulich. – Wie schwer die Verdauungsarbeit ist, zeigt uns die Tatsache, daß wir nach dem Essen müde werden und sowohl geistig als auch körperlich zu keinen Höchstleistungen unmittelbar nach dem Essen fähig sind. – Dies bedeutet also, daß wir dem Körper insbesondere von Seiten des Herzens und Kreislaufs sowie des Darms, der Leber, Nieren und übrigen inneren Organe durch diese Kost entlasten.

Die Vorteile dieser Astronautennahrung sind also:

a) daß die Kraft, Widerstandsfähigkeit und der Heilprozeß gesteigert werden, indem dem Körper all das, was er braucht, ausreichend und richtig zur Verfügung gestellt wird,

b) daß der Körper von unnötigen Anstrengungen entlastet wird,

c) daß wenig Stuhl anfällt, da diese Nahrung keine Ballaststoffe enthält, das Ausbleiben von Stuhlgang ist also kein Zeichen von Verstopfung, sondern ein Zeichen der nicht vorhandenen Masse sprich Ballaststoffe, – wodurch Sie in den ersten Tagen nach der Operation von dem unangenehmen Problem des „Töpchensitzens" befreit werden,

d) daß die Wundheilung erheblich verbessert wird, weil die Gefahr der Keimverschleppung durch den ausbleibenden Stuhlgang unterbrochen wird,

e) daß zusammen mit unserer sehr schonenden Anaesthesietechnik bereits unmittelbar nach der Operation erlaubt ist, soweit Sie Hunger oder Durst verspüren, diese „Astronautennahrung" und „Drinks" in unbegrenzter Menge zu sich zu nehmen.

Sie sollten aber darauf achten, daß Sie täglich, d. h. über 24 Stunden, wenigstens 1,5 Liter dieser Nahrung und Drinks insgesamt zu sich nehmen. Überschreiten können Sie die Menge dieser Ernährung soweit es Ihre Figur erlaubt. Andere Nahrung oder Getränke dürfen Sie allerdings nicht zu sich nehmen, da Sie damit den sorgfältigen Aufbau dieser Nahrung zerstören und die gesamte Ernährung zur Entgleisung bringen mit Durchfällen usw.

Wie alles im Leben hat auch diese Nahrung mit all ihren Vorteilen eine zweite Seite. Da sie leicht verdaulich und arm bzw. frei von Ballaststoffen ist, ausreichend Flüssigkeit, Salze usw. enthält, muß sie so flüssig sein wie sie ist.

Diese „Astronautennahrung" mit „Mineraldrinks" erhalten Sie am Tage der Aufnahme und am Operationstag auf eigenen Wunsch oder auf Vorschlag von uns. Danach bekommen Sie wieder die Normalkost. Prüfen Sie aber bitte zuvor, ob Sie nicht doch übergewichtig sind und deshalb auf die von dem Haus angebotene Diät zum „Abspecken" anstelle der Normalkost übergehen sollten.

Bestimmt wissen Sie schon, daß in der Bundesrepublik Deutschland mehr als die Hälfte aller Vierzigjährigen übergewichtig sind und daß damit der Wohlstandsspeck zu einem wahrhaft schwerwiegenden Problem für die Gesunderhaltung geworden ist.

Fettleibigkeit ist Mitursache für viele Erkrankungen. Schon ein Übergewicht von einigen Kilogramm kann Ihre Lebenserwartung verringern. Eines der wichtigsten Mittel für die Erhaltung Ihrer Gesundheit ist daher – die Waage. Kontrollieren Sie damit täglich Ihr Gewicht.

Fig. 1 (contd.)   *(will be continued)*

So ermitteln Sie Ihr statistisches Normalgewicht:

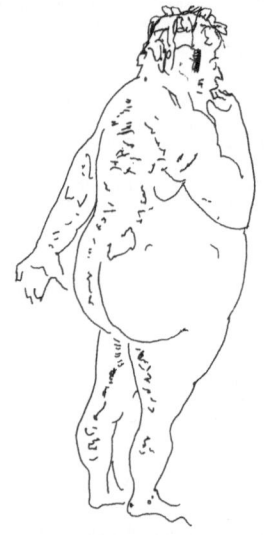 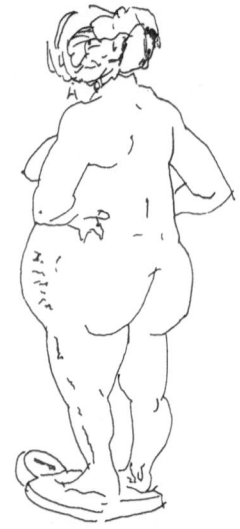

Männer	
Größe in cm:	170 cm
davon abziehen:	100
Normalgewicht =	**70 kg**

Frauen	
Größe in cm:	170 cm
davon abziehen:	100
	70 kg
davon außerdem abziehen 10 Prozent:	7
Normalgewicht =	**63 kg**

Und denken Sie daran: Ihr medizinisches Idealgewicht liegt noch unter diesen Gewichtsangaben.

Sie sollten wissen,

- daß wir – im Durchschnitt – täglich 300 bis 500 Kalorien mehr essen und trinken als wir verbrauchen;

- daß wir – im Durchschnitt – heute je zehn Kilogramm mehr wiegen als unsere Vorfahren vor einem Jahrhundert;

- daß zu hoher Blutdruck bei Übergewichtigen etwa dreimal häufiger vorkommt als bei Normalgewichtigen;

- daß jeder Tag, den Sie mit erhöhtem Blutdruck verbringen, Ihr Leben verkürzen kann;

- daß der zur Volkskrankheit gewordene Diabetes (Zuckerkrankheit) und die an erster Stelle der Sterblichkeitsstatistik stehenden Herz-Kreislauf-Erkrankungen bei Übergewichtigen wesentlich häufiger vorkommen als bei Normalgewichtigen;

- daß Übergewicht eine schwere Belastung für sämtliche Gelenke, von den Beinen über die Hüften bis zur Wirbelsäule ist und bereits verschlissene oder verletzte Gelenke hierdurch noch schneller weiter zugrunde gerichtet werden.

Hinzu kommt die erhebliche Belastung Ihres Herzens. Denn das gesamte Fettgewebe muß und wird dauernd von Blut durchströmt. Diese Arbeit muß in jedem Augenblick, tagaus, tagein über 24 Stunden von dem Herzen geleistet werden, d. h., daß Ihr Herz alleine schon deshalb auch während der Ruhezeit mit mehr oder weniger „Vollgas", also mit Hochleistung arbeiten muß. Es muß und wird sich darum, wie jeder mit Hochleistung fahrende Motor, früher verschleißen. Wird Ihrem Herzen aber gar eine noch etwas höhere Leistung im beruflichen Leben oder Ihrem Haushalt abverlangt, so wird es sehr bald mit seiner Höchstleistung antworten oder sogar die Höchstleistung überschreiten müssen, um einmal dieses Übergewicht dauernd transportieren und zu allem Überfluß auch noch ständig mit Blut durchspülen zu können.

Fig. 1 (contd.)   *(will be continued)*

Darüber hinaus werden die Leber, die Bauchspeicheldrüse und die übrigen Verdauungsorgane in ihrer Tätigkeit durch die Verfettung behindert, da die Zellen durch Fettgewebe in ihrer Funktion beeinträchtigt und eingeengt werden. Dies führt u. a. zur Zuckerkrankheit, zum Bluthochdruck mit Kopfschmerzen, körperlicher Abgeschlagenheit, Müdigkeit, zu Leber- und Gallenerkrankungen usw. Die Fettleibigkeit führt auch zu einer Verdrängung des Zwerchfells in die Brusthöhle und damit zu einer Einschränkung der Atmung mit Kurzatmigkeit, Lufthunger, körperlicher und geistiger Trägheit sowie Leistungsunvermögen.

Um diesen sich ständig steigernden Teufelskreis aufzuhalten und abzubauen, müssen Sie Ihr Körpergewicht verringern! Dies erreichen Sie, indem Sie die angehäuften Vorräte, die unsere Fettpolster darstellen, verbrauchen d. h., Sie müssen von Ihrem Körper verbrannt werden, indem Sie gleichzeitig die Zufuhr solcher fettpolstererzeugender Stoffe stoppen, die Sie ja schon im Überfluß haben.

Das Krankenhaus für Sportverletzte hat für Sie vom Frühstück bis zum Abendessen eine spezielle Diät zusammengestellt, die aus **Fleisch, Fisch, Käse, Eiern und Salaten** besteht. Mit dieser eiweiß- und vitaminreichen Kost können Sie wöchentlich wenigstens 1 kg Fett loswerden. Bitte denken Sie aber daran, daß jegliches zusätzliches Essen absurd ist und die Zeit Ihres Aufenthaltes bei uns in der Klinik nur zur Anleitung für Sie dienen kann. Die eigentliche Aufgabe muß aber für Sie die Weiterführung dieser begonnenen Ernährungsumstellung zu Hause bleiben, wobei ganz besonders Zucker, auch in seinen abgewandelten Formen wie Bier, Kuchen, Schokolade, Alkoholika, Nudeln, Reis, Säfte usw. völlig von der Speisekarte ausgeschlossen bleibt.

Wichtig ist auch, daß Sie **1,5 Ltr. kalorienfreie Flüssigkeit** wie einfaches Wasser, Sprudelwasser, Tee oder Kaffee ohne Zucker, mit Wasser verdünntes, künstlich gesüßtes Zitronenwasser oder dergleichen zu sich nehmen. - Bitte während Ihres Aufenthaltes in unserer Klinik täglich bei der Stationsschwester verlangen.- Sie müssen diese Menge Flüssigkeit täglich aus vielerlei Gründen trinken. Einer der wichtigsten Gründe ist, daß der Kreislauf nur auf diese Weise stabil bleiben kann. Sie vermeiden dadurch u. a. auch größere Hungergefühle. Auf keinen Fall werden Sie dadurch dicker oder in Ihrer Gewichtsabnahme behindert, im Gegenteil.

Während der Abmagerungsdiät kommt es gelegentlich zu „Schwarzwerden vor den Augen", leichtem Wiegen oder Karussellfahren des Bettes" und „Verlangen häufiger tief durchzuatmen". Dies sind alles Zeichen dafür, daß der Körper das Körperfett verbraucht oder, wie wir Mediziner sagen, „das Fett verbrennt". Für diese Fettverbrennung braucht der Körper mehr Sauerstoff als z. B. für die normale Zuckerverbrennung. Dies wiederum ist die Ursache für das Bedürfnis, häufiger tief durchzuatmen, um dem Körper den notwendigen Sauerstoff hierfür zuzuführen. Die Fettverbrennung und Zuckerverbrennung im menschlichen Körper ist vergleichbar mit der Verbrennung von Dieselkraftstoff und Benzin. Der Dieselkraftstoff verbraucht mehr Sauerstoff für einen Verbrennungsvorgang als der Benzinkraftstoff und ebenso ist der Mensch während der Fettverbrennung auch „nur so schnell wie ein Dieselmotor". Diese Schwerfälligkeit der Fettverbrennung und damit der Bereitstellung von Energie für unser Leben ist auch der Grund für das „Schwarzwerden vor den Augen"oder „Karussellfahren". Wenn diese Symptome also auftreten, dann trinken Sie nur etwas mehr von der kalorienfreien Flüssigkeit (Wasser) und freuen sich im übrigen darüber, weil Sie jetzt wirklich beginnen, Ihren Wohlstandsspeck loszuwerden. Die von uns für Sie zusammengestellte Abmagerungsdiät berücksichtigt, daß Ihnen all das, was Sie an Nahrungsstoffen während des Abmagerns brauchen, sogar in Überschuß zugeführt wird. Sie gehen damit also kein Risiko ein.

Um keine irrigen Vorstellungen aufkommen zu lassen, muß an dieser Stelle noch erwähnt werden, daß im Grundpflegesatz aller Patienten für das Essen ein Tagessatz von nur 8,50 DM enthalten ist. Der gesamte Restbetrag des Grundpflegesatzes ist voll für die Krankenhausvorhaltekosten mit Personalkosten usw. erforderlich!!

Nicht nur die Speisen und Getränke, sondern auch die Eßgewohnheiten müssen umgestellt werden. Wichtigster Grundsatz ist, daß Sie morgens zu Ihrer gewohnten Zeit in Ruhe frühstücken. Hastiges Essen oder gar Schlingen ist der erste Todfeind Ihrer Figur und Gesundheit.

Kontrollieren Sie daher konsequent Ihr Eßtempo mit der Uhr. 1 Minute Kauen muß durch wenigstens 15 Sekunden Pause unterbrochen werden. Keine Frühstücksgespräche! Sie verführen ebenso wie die Zeitung zum gedankenlosen Vielessen.

Führen Sie Kauen und Trinken nicht zugleich durch. Es sind zwei getrennte Vorgänge, die man ebenso wenig wie Schlafen und Wachen oder Weinen und Lachen vereinen soll.
Zum Frühstück lassen Sie sich wenigstens 15 Minuten und zum Mittag- sowie Abendessen wenigstens 30 Minuten Zeit.

Fig. 1 (contd.)   *(will be continued)*

195

Genießen Sie Ihr Essen ! Dies können Sie nur, wenn Sie bedächtig, gut und ohne Ablenkung kauen. Denken Sie bitte immer daran, daß sich alle anderen Fleischesser wie Löwen, Tiger, Leoparden usw. für eine einzige Mahlzeit viele Stunden Zeit gönnen und dabei sehr langsam, bedächtig und voller Genuß ihre Beute über Tage verteilt verzehren oder aber, daß die „Schlinger" wie z. B. Schlangen, sich zum Verdauen Monate Zeit lassen.

Der Erfolg muß natürlich auch durch tägliches Wiegen am Morgen vor dem Frühstück, nach Entleeren von Darm und Blase,kontrolliert werden.

Die Kalorien für 100 g oder 100 ccm der folgenden Nahrungsmittel sind:

Verboten: Fette – 900, Wurst – 524, Nüsse – 561-687, Schokolade – 520, Hülsenfrüchte– 348, Datteln, Feigen, Rosinen – 289, Brantwein 280, Brot – 269-389, Spaghetti – 369, Reis – 360, Kartoffelchips – 568, Limonaden, Bier – 48-66, Wein – 60-120 cal, Mais – 96, Bananen – 85, Apfelsinen, Birnen – 60, Kartoffeln – 76.

Erlaubt: Gurken – 13, Salat – 14, Rhabarber – 16, Sellerie – 17, Radieschen, Sauerkraut – 18, Spargel – 21, Tomaten, Pilze – 22, Paprika grün – 24, Spinat, Rotkohl, Mangold – 27, Kohlrabi – 29, Wirsing – 31, Grünkohl – 38, Lauch – 44.

Notwendig: Fleisch, Fisch, Käse, Eier.

Nach diesem Ernährungsexkurs aber zurück zur Anaesthesieausleitung.

Die erste Zeit nach der Operation verbringen Sie in einem besonderen „Aufwachraum" unter der Obhut speziell geschulten Pflegepersonals. Hier können Sie auch einen Becher Flüssigkeit gegen Ihren Durst verlangen. Sobald Sie wieder ausreichend „Herr" bzw. „Frau" Ihrer selbst sind, werden Sie auf Ihr Zimmer zurückgebracht.

Nach einem operativen Eingriff kann auch einmal eine intensivmedizinische Versorgung erforderlich werden, wofür wir mit einer Spezial-Abteilung vorgesorgt haben. Hier können wir Behandlungen bis zur künstlichen Beatmung, Ernährung, Unterhaltung von Herzkreislauffunktionen und Unterkühlung durchführen. Diese Einrichtung ist allerdings mit nichts mehr in der Fliegerei vergleichbar (!), denn hier am Ziel angekommen, werden Sie aus der Obhut der Fluggesellschaft entlassen; bei uns werden Sie immer noch mit einem großen unsichtbaren Sicherheitsaufwand beschützt.

Das Sicherheitsrisiko für Sie im Straßenverkehr, im Haushalt, im Betrieb, auf Reisen (mit Schiff, Flugzeug, Eisenbahn), bei Sport und Vergnügen und durch Gewaltverbrechen, kurzum im gesamten täglichen Leben liegt unvergleichlich weit über dem Risiko der Anaesthesie in unserem Hause. Man kann ohne Übertreibung sagen, daß Sie während der Anaesthesie bei uns, vom Einschlafen bis zum Wiederaufwachen, in sicherer Obhut sind als in der Gesamtlebenssituation unter Ihrer eigenen Verantwortung.

Hierfür noch einige statistische Daten: Im normalen Alltag des Jahres 1973 verunglückten in der Bundesrepublik 35100 Menschen tödlich, anders ausgedrückt, es starben jeder 1794ste Bundesbürger oder 0,06% aller Bundesbürger / Jahr durch Gewalteinwirkung. Durch den Tabak sterben z. Zt. sogar viermal soviel Menschen, also rund 0,24%.

Die Anaesthesiestatistiken anderer Kliniken weisen eine Rate von 0,3-0,07% auf ihre Patienten aus.

Unsere Gesamtstatistik liegt erfreulicherweise heute bei 0,0000%, wobei wir alleine an unserer Klinik jährlich 5000 Anaesthesien durchführen.

Dennoch bleibt jede Anaesthesie ein gewisses Risiko und wir sind daher ständig bemüht, unsere Arbeit noch sicherer und angenehmer für Sie zu gestalten. Dabei sind wir auf Ihre Unterstützung angewiesen. Einen Tag nach der Narkose werden Sie einen postanaesthesialogischen Patientenfragebogen erhalten. Wir bitten Sie, anhand dieses Fragebogens Ihre Eindrücke zur abgelaufenen Narkose zu schildern. Um Ihre Angaben für unsere praktische Arbeit verwerten zu können, ist es unbedingt erforderlich, daß Sie **a l l e** aufgeführten Fragen nach bestem Wissen beantworten, auch wenn Ihnen die Bedeutung im Einzelfall nicht ohne weiteres überzeugend erscheint. Positive wie negative Kritik sind hierin eingeschlossen.

– So bieten wir Ihnen heute aufgrund dieser Befragung schon nach Erwachen aus der Narkose ein Getränk an, weil sich viele Ihrer Vorgänger über einen trockenen Mund, Durstgefühl, etc. beklagt haben.–

Fig. 1 (contd.)   *(will be continued)*

Zum Abschluß noch folgende **WICHTIGE HINWEISE:**

Am Abend vor der Operation dürfen Sie nach dem Abendbrot **nichts mehr essen**

nach 22.00 Uhr auch **nichts mehr trinken**

sowie beim Zähneputzen am Operationsmorgen **kein Wasser schlucken**

am Tage vor der Operation und am Operationstag sollen Sie auch **nicht rauchen**

auf dem Zimmer müssen bleiben **Zahnprothesen, Brille, Schmuck, Ehering u.s.w.**

und entfernen Sie am Abend vor der Operation **Schminke (Make up), Hautcreme, Nagellack u.s.w.**

am Operationstag sind Sie bitte um **7.00 Uhr**
mit der Morgentoilette fertig, falls Sie von der Stationsschwester keine andere Anweisung erhalten haben, da Sie dann Ihre Beruhigungsspritze bekommen und danach nicht mehr aufstehen dürfen.

Sollten Sie vor der Operation noch etwas warten müssen, was gelegentlich leider nicht zu vermeiden ist, so bitten wir Sie jetzt schon um Ihr Verständnis dafür. Eine Operation und Anaesthesie mit all den notwendigen und aufwendigen Vor- und Nachuntersuchungen sowie Behandlungen kann immer mit unvorhersehbaren Überraschungen verbunden sein, die ganz plötzlich zusätzliche Zeit erfordern. Dies kann sowohl für Sie als auch die Patienten vor oder nach Ihnen der Fall sein. Haben Sie darum für eine solche Situation, auch wenn sie einen Patienten vor Ihnen betrifft, Verständnis und Geduld, denn wir alle wollen nur sehr gute und sehr sorgfältige Arbeit für unsere Patienten leisten.

Um Ihnen die Wartezeit vor der Operation zu verkürzen, erleichtern und auch verschönern zu können, bieten wir Ihnen ein Musikprogramm an. Bei der Zusammenstellung dieses Programms haben wir uns bemüht, Ihren Wünschen und den Wünschen anderer Patienten gerecht zu werden. Sie werden verstehen, daß wir dabei spezielle Wünsche nach bestimmten Interpreten oder Musiktiteln nicht erfüllen können. Dennoch wird Sie die angebotene Musik sicherlich unterhalten, ablenken und auch inspiriren können.

Versuchen Sie beim anhören der Musik nicht an Ihre Operation, sondern an irgend etwas Schönes oder Interessantes zu denken.

Wir würden uns jedenfalls freuen, wenn es uns gelänge, Ihnen auf diese Weise die Zeit vor der Operation angenehmer zu gestalten!

Fig. 1 (contd.) *(will be continued)*

197

Mindestens 6 Stunden **v o r d e r** Operation dürfen Sie

nichts mehr essen

nichts mehr trinken

nicht mehr rauchen

Wenn Sie etwas zu sich genommen haben,

teilen Sie dies dem Arzt mit!

Vor der Operation entfernen Sie bitte

Zahnprothese, Brille, Schmuck, Ehering u.s.w.

sowie

Schminke (Make up), Hautcreme, Nagellack u.s.w.

Mindestens 24 Stunden

nach der Operation

dürfen Sie auch als Fußgänger

nicht am Straßenverkehr teilnehmen

sondern müssen sich vielmehr der Führung anderer Personen anvertrauen.

Sie sollten außerdem

keine wichtigen Entscheidungen treffen

keine Maschinen oder Geräte bedienen

und

keinen Alkohol zu sich nehmen.

Fig. 1 (contd.)   *(will be continued)*

gelesen am: ....

Erklärung des Patienten zum Aufklärungsgespräch:

Herr Dr. . . . . hat heute mit mir anhand meiner Antworten ein Gespräch über das Anaesthesieverfahren geführt. Ich konnte alle mich interessierenden Fragen, insbesondere nach der Art des Verfahrens und seinen spezifischen Risiken, nach der Vor- und Nachbehandlung sowie etwaigen Nebeneingriffen stellen. Ich habe keine weiteren Fragen.

Fig. 1 (contd.)

A N A E S T H E S I O L O G I E

Sportkrankenhaus Hellersen
Chefarzt : Dr. R. Droh

Krankenfragebogen

Adressette (bzw. Name [bei Frauen auch Mädchenname], Vorname, Geb.-Datum, Adresse)

Sehr verehrte Patientin
Sehr geehrter Patient

Bitte füllen Sie diesen Fragebogen sofort nach Erhalt sorgfältig aus. Damit erleichtern Sie uns die Arbeit und Sie selbst haben auch die Möglichkeit, länger als bei einer direkten Befragung über die eine oder andere Frage nachzudenken. Durch Ihre Mitarbeit helfen Sie uns aber auch, das Beste für Ihre Sicherheit zu tun! Wundern Sie sich bitte nicht über die Zahl der Fragen, sondern bedenken Sie vielmehr, daß der Anaesthesist für die Schmerzfreiheit während der Operation und die Überwachung und Aufrechterhaltung der lebenswichtigen Funktionen (z. B. Atmung, Herz- und Kreislauf usw.) zu sorgen hat. Ihre Antworten sollen dem Anaesthesisten also ein Bild über Ihren Gesundheitszustand vermitteln. Jede Frage bezieht sich auf bestimmte, z. T. auch geringfügige Risiken (z. B. Zahnschäden, Venenreizungen u. s. w.).

Diesen Fragebogen händigen Sie bitte ausgefüllt Ihrer Stationsschwester vor der Untersuchung aus. Fragen Sie den Anaesthesisten bei der Vorstellung bzw. bei seiner Visite, die auch Ihrer Aufklärung dienen soll, nach allem, was Sie im Zusammenhang mit der Anaesthesie interessiert.

Unterschreiben Sie bitte nach diesem Aufklärungsgespräch mit dem Arzt die Erklärung am Schluß dieses Fragebogens.

Bitte beantworten Sie nun die nachstehenden Fragen, indem Sie das zutreffende ☐ mit einem X (Kreuz) versehen.

Zur Familie :	wenn lebend Alter und Gesundheitszustand	wenn verstorben Sterbealter und Todesursache
Vater :		
Mutter :		
Geschwister :		

Litt jemand Ihrer nächsten <u>Verwandten</u> an :

Lungen- oder anderer Tuberkulose ☐ Ja ☐ Nein
Zuckerkrankheit . . . . ☐ Ja ☐ Nein
Bluthochdruck . . . . ☐ Ja ☐ Nein
Schlaganfall . . . . ☐ Ja ☐ Nein
Asthma . . . . ☐ Ja ☐ Nein
Nervenkrankheit . . . . ☐ Ja ☐ Nein
Muskelerkrankungen . . . . . . ☐ Ja ☐ Nein
Narkosezwischenfall m. Überhitzung ☐ Ja ☐ Nein
Muskelschmerzen n. Alkoholgenuß ☐ Ja ☐ Nein

Haben Sie <u>selbst</u> Kinderkrankheiten durchgemacht und welche?

Masern . . . . ☐ Ja ☐ Nein
Keuchhusten . . . . ☐ Ja ☐ Nein
Scharlach . . . . ☐ Ja ☐ Nein
Windpocken . . . . ☐ Ja ☐ Nein
Diphterie . . . . ☐ Ja ☐ Nein
Röteln . . . . ☐ Ja ☐ Nein
Polio (Kinderlähmung) . . . . . ☐ Ja ☐ Nein
Haben Sie selbst starke Muskelschmerzen schon nach geringem Alkoholgenuß . . ☐ Ja ☐ Nein

Fig. 2   *(will be continued)*

Sind Sie schwerhörig?	□ Ja	□ Nein

beiderseits □   rechts □,   links □

Haben Sie lose Zähne ?	□ Ja	□ Nein
Haben Sie eine Zahnvollprothese des Unterkiefers ?	□ Ja	□ Nein
Haben Sie eine Zahnteilprothese des Unterkiefers ?	□ Ja	□ Nein
Haben Sie eine Zahnvollprothese des Oberkiefers ?	□ Ja	□ Nein
Haben Sie eine Zahnteilprothese des Oberkiefers ?	□ Ja	□ nein

Waren Sie einmal ernsthaft krank, abgesehen von Ihrer jetzigen Erkrankung:

am Herzen ?	□ Ja	□ Nein

(z. B. Herzmuskelschwäche, Klappenfehler, Infarkt, Durchblutungsstörungen der Herzkranzgefäße, Herzbeklemmungen, Angina pectoris, Herzmuskelentzündung)

wenn ja von ? _____ bis _____

Wurde ein Herzschrittmacher bei Ihnen implantiert ?	□ Ja	□ Nein
am Kreislauf ? (z. B. Schwindel, erhöhter oder erniedrigter Blutdruck)	□ Ja	□ Nein

wenn ja, von _____ bis _____

Haben Sie abends regelmäßig dicke Beine ?	□ Ja	□ Nein
Haben Sie Durchblutungsstörungen ?	□ Ja	□ Nein
Haben Sie Krampfadern ?	□ Ja	□ Nein
Hatten Sie eine Thrombose ?	□ Ja	□ Nein

wenn ja, wann ? _____

an der Lunge und den Atemwegen ?	□ Ja	□ Nein

(z. B. Tuberkolose, Staublunge, Lungenentzündung, Lungenblutung, Rippenfellentzündung, Bronchitis)

wenn ja, von _____ bis _____

Asthma ?	□ Ja	□ Nein

wenn ja, von _____ bis _____

Atemnot bei Belastung wie Treppensteigen ?	□ Ja	□ Nein

wenn ja, von _____ bis _____

Atemnot ohne Belastung ?	□ Ja	□ Nein
an den Nieren ? (z. B. Nierensteine, Nierenentzündung)	□ Ja	□ Nein

wenn ja, von _____ bis _____

Ist Ihr Urin auffällig verändert ?	□ Ja	□ Nein

wenn ja, seit wann ? _____

an der Leber ? (z. B. Leberentzündung, Gallenblasenentzündung, Fettleber, Leberverhärtung)	□ Ja	□ Nein

wenn ja, von _____ bis _____

Gelbsucht ?	□ Ja	□ Nein

wenn ja, von _____ bis _____

Hatten Sie in den letzten 8 Wochen wandernde Schmerzen in Hand- oder Fußgelenken oder der Wirbelsäule mit leichtem Fieber?.	□ Ja	□ Nein
Bestanden in den letzten 8 Wochen Appetitlosigkeit, Widerwillen gegen Fette und Übelkeit ?	□ Ja	□ Nein
Ist Ihr Stuhlgang auffällig verändert ?	□ Ja	□ Nein

wenn ja, seit wann ? _____

an Zuckerkrankheit ?.	□ Ja	□ Nein

wenn ja, seit wann ? _____

wie eingestellt ? _____

Haben Sie vermehrt Durst ?	□ Ja	□ Nein
an Geschwulsten ?	□ Ja	□ Nein
an Fettstoffwechselstörungen ?	□ Ja	□ Nein
an Wirbelsäulenerkrankungen ?	□ Ja	□ Nein
an Muskelschwäche ?	□ Ja	□ Nein
an Gelenkverschleiß ?	□ Ja	□ Nein

Fig. 2 (contd.)   *(will be continued)*

an Gicht ?	☐ Ja	☐ Nein

an Gicht ? . . . . ☐ Ja ☐ Nein

an Rheumatismus ? . . . . ☐ Ja ☐ Nein

am Hals ? . . . . ☐ Ja ☐ Nein

an den Mandeln ? . . . . ☐ Ja ☐ Nein

an der Schilddrüse (Kropf) ? . . . . ☐ Ja ☐ Nein

wenn ja, haben Sie ein Schilddrüsenpräparat eingenommen ? . . . . . ☐ Ja ☐ Nein

wenn ja, welches, bitte Namen nennen ? _____

an Bluterkrankungen ? (z. B. Blutarmut, Leukämie) . . . . ☐ Ja ☐ Nein

an Drüsenerkrankungen ? (z. B. Bauchspeicheldrüse, Brustdrüse) . . . ☐ Ja ☐ Nein

an Magen- Darmerkrankungen ? (z. B. Durchfälle) . . . . . ☐ Ja ☐ Nein

am Nervensystem ? (z. B. Krämpfe, Lähmungen, Gehirnerschütterung) . . ☐ Ja ☐ Nein

an Gemütsleiden ? (z. B. Depressionen) . . . . ☐ Ja ☐ Nein

Haben Sie im Laufe des letzten Jahres ein Nebennierenrindenpräparat,
(z. B. Solu-Decortin H, Urbason, Volon) eingenommen ? . . . . . ☐ Ja ☐ Nein

wenn ja, welches ? _____

Leiden sie an Porphyrie ? (spez. Erkrankung) . . . . ☐ Ja ☐ Nein

Ist bei Ihnen eine Überempfindlichkeit (Allergie) bekannt, gegen:

Jod ☐ Ja ☐ Nein   Medikamente ☐ Ja ☐ Nein   Primel ☐ Ja ☐ Nein

Pflaster ☐ Ja ☐ Nein   Waschmittel ☐ Ja ☐ Nein   Erdbeeren ☐ Ja ☐ Nein

oder eine andere Substanz (wenn ja, bitte beschreiben) ? _____

Nehmen Sie irgendwelche Medikamente dauernd ein ? . . . . . . ☐ Ja ☐ Nein

wenn ja, welche ? _____

in welcher Dosierung ? _____

Nehmen Sie häufig Schlafmittel oder Betäubungsmittel ? . . . . . ☐ Ja ☐ Nein

wenn ja, welche ? _____

Nehmen Sie laufend Mittel gegen Darmträgheit ? . . . . ☐ Ja ☐ Nein

Wurde Ihnen schon einmal Blut übertragen ? . . . . ☐ Ja ☐ Nein

wenn ja, sind dabei Unverträglichkeiten aufgetreten ? . . . . . . ☐ Ja ☐ Nein

Hatten Sie schon einmal eine Vollnarkose ? . . . . ☐ Ja ☐ Nein

wenn ja, wann ? _____

und wissen Sie welches Narkosemittel verwendet wurde ? _____

Sind irgendwelche Narkoseunverträglichkeiten bekannt ? . . . . . ☐ Ja ☐ Nein

wenn ja, welche ? _____

wurden oder werden Sie bestrahlt ? . . . . ☐ Ja ☐ Nein

Besteht bei Ihnen eine Schwangerschaft ? . . . . ☐ Ja ☐ Nein

Besteht bei Ihnen eine Regelstörung ? . . . . ☐ Ja ☐ Nein

wenn ja, wann ? _____

Besteht bei Ihnen eine Unterleibserkrankung ? . . . . ☐ Ja ☐ Nein

Ist die Geburt Ihrer Kinder normal verlaufen ? . . . . ☐ Ja ☐ Nein

wenn nein, welche Schwierigkeiten gab es ? _____

Haben Sie eine Abneigung gegen irgendwelche Speisen ? . . . . . ☐ Ja ☐ Nein

wenn ja, gegen welche ? _____

Trinken Sie regelmäßig Alkohol ? . . . . ☐ Ja ☐ Nein

wenn ja, in welcher Form und wieviel pro Tag ? _____

Rauchen Sie ? . . . . ☐ Ja ☐ Nein

wenn ja, wieviel pro Tag ? _____ seit wann _____

Schlafen Sie schlecht ? . . . . ☐ Ja ☐ Nein

Haben Sie ein Glaukom (grüner Star) ? . . . . ☐ Ja ☐ Nein

Fig. 2 (contd.)   *(will be continued)*

Ermüden Sie schnell, z, B, beim <u>Kämmen</u> Ihrer Haare ? . . . . . . □ Ja □ Nein

Gehören Sie einer Glaubensgemeinschaft an, die jegliche Übertragung von Blut
oder einzelner Bestandteile verbietet ? . . . . □ Ja □ Nein

<u>Bluten</u> Sie nach Verletzungen besonders stark oder lang ? . . . . . □ Ja □ Nein

Haben Sie eine schlechte Wundheilung ? . . . □ Ja □ Nein

Neigen Sie zu Blutergüssen, Nasenbluten etc. ? . . . . . . □ Ja □ Nein

Neigen Sie besonders leicht zum <u>Erbrechen</u> ? . . . . . □ Ja □ Nein

Nehmen Sie <u>Drogen</u> (Rauschgift) ? . . . . □ Ja □ Nein

Haben Sie Schwierigkeiten beim Rückwärtsbeugen des Kopfes ? . . . □ Ja □ Nein

Haben Sie Schwierigkeiten den Mund weit zu öffnen ? . . . . □ Ja □ Nein

Waren Sie an anderen Organen erkrankt ? . . . . . . □ Ja □ Nein

wenn ja, welches Organ ? _____

von _____ bis _____

Befanden Sie sich in letzter Zeit in ärztlicher Behandlung ? . . . . □ Ja □ Nein

wenn ja, wegen welcher Erkrankung ? _____

Wurden Sie in den letzten 6 Wochen geimpft ? . . . . . □ Ja □ Nein

wenn ja, wogegen ? _____

Leiden Sie an einer anderen, oben nicht aufgeführten Erkrankung ? . . . □ Ja □ Nein

wenn ja, welcher ? _____

Haben Sie noch irgendwelche weitere Beschwerden ? . . . . . □ Ja □ Nein

wenn ja, welche ? _____

seit wann ? _____

Fühlen Sie sich unabhängig von Ihrem zu operierenden Leiden krank ? . . . □ Ja □ Nein

Treiben Sie Sport ? . . . . □ Ja □ Nein

Haben Sie Sport getrieben ? . . . . . . □ Ja □ Nein

wenn ja, welche Sportart ? _____

in welchem Verein ? _____

wurden Sie schon einmal im <u>Sportkrankenhaus Hellersen</u> operiert ? .. . . □ Ja □ Nein

wenn ja, warum ? _____

wann ? _____

Wurden Sie schon einmal in einem auswärtigen Krankenhaus operiert ? . .. . □ Ja □ Nein

wenn ja, weshalb ? _____

wo ? _____

wann ? _____

Wer ist Ihr Hausarzt ? _____

Wünschen Sie am Morgen vor der Operation zur Überbrückung der Wartezeit Musik ? . □ Ja □ Nein

Wenn ja, haben Sie einen spez. Wunsch ? _____

_____

Welche Musik bevorzugen Sie ? _____

Klassik □,   Moderne □,   Militär □,   Schlager □ ? _____

_____

Sonstige Bemerkungen: _____

_____

Ich bestätige hiermit die Richtigkeit und Vollständigkeit meiner Angaben und erkläre entsprechend aufge-
klärt worden zu sein, weiter erkläre ich mich mit der Durchführung einer Allgemeinnarkose bzw. Lokal-
anaesthesie einverstanden.

_____   _____
       Datum   Unterschrift des Patienten
                                       bzw. seiner Erziehungsberechtigten

Fig. 2 (contd.)

A N A E S T H E S I O L O G I E

Sportkrankenhaus Hellersen
Chefarzt: Dr. R. Droh

5880 Lüdenscheid
Tel.: (02351) 434218-9

Betr.:

Adressette (bzw. Name [bei Frauen auch Mädchenname], Vorname Geb.-Datum, Adresse)

Herrn / Frau (Kollegen/in)

Zur Abkürzung der Informationswege bitte Anforderungen, Laborbefunde oder Konsiliarvorschläge direkt auf dieses standardisierte Formblatt übertragen und   Zutreffendes bitte unterstreichen bzw. ankreuzen:

☐ telefonisch durchzugeben
☐ zurückzusenden
☐ der / dem Patienten(in) mitzugeben

Sehr verehrte Frau Kollegin!
Sehr geehrter Herr Kollege!

Die / der Patient/in wird/ wurde am _____ operiert / anaesthesiologisch voruntersucht.
In diesem Zusammenhang werden / wurden die umseitig angekreuzten Daten erforderlich / festgestellt.

Aufgrund der Befunde müssen / mußten folgende / keine praeoperativen Maßnahmen getroffen werden:
1.
2.
3.
Die Operation wird / wurde dadurch um _____ Tage / nicht verschoben.

Während der Anaesthesie und Operation ergeben / ergaben sich folgende / keine Besonderheiten.

Postoperativ   werden   keine   weiteren / folgende   diagnostischen   / therapeutischen   Maßnahmen   vorgeschlagen:

Fig. 3

(will be continued)

204

Klinische Besonderheiten:

Anforderung = X̸

Urinwerte

Biogramm

Urinkultur ☐

Blutkultur ☐

Keimzahl:

Keime

		Normwerte	Datum	Datum	Datum	Datum	Datum
Urinstatus	R / pH		☐	☐	☐	☐	☐
Nitrit		negativ	☐	☐	☐	☐	☐
Blut		negativ	☐	☐	☐	☐	☐
Zucker	Z %	negativ	☐	☐	☐	☐	☐
Aceton	Ac.	negativ	☐	☐	☐	☐	☐
Eiweiß	Eiw. Esb. %	negativ	☐	☐	☐	☐	☐
Sed.	Ery.	1-2/BF.	☐	☐	☐	☐	☐
	Leuko.	1-5/BF.	☐	☐	☐	☐	☐
Rundepith.	negativ		☐	☐	☐	☐	☐
Plattenepith.	(+) – +		☐	☐	☐	☐	☐
	Zyl.	negativ	☐	☐	☐	☐	☐
	Trichomonaden	negativ	☐	☐	☐	☐	☐
	Bakt.	negativ	☐	☐	☐	☐	☐
	Salze		☐	☐	☐	☐	☐
	Schleim		☐	☐	☐	☐	☐
Sammelurin	Menge l/24 H		☐	☐	☐	☐	☐
	Zucker g/24 h		☐	☐	☐	☐	☐
Natrium	Na⁺ mval/l	100-180	☐	☐	☐	☐	☐
Kalium	K⁺ mval/l	60-90	☐	☐	☐	☐	☐
Calcium	Ca⁺⁺ mval/l	5-15	☐	☐	☐	☐	☐
Chlorid	Cl mval/l	100-200	☐	☐	☐	☐	☐
Gallenfarbst.	U'gen	negativ	☐	☐	☐	☐	☐
	Urob.	negativ	☐	☐	☐	☐	☐
	Bil.	negativ	☐	☐	☐	☐	☐
Diazoreaktion	Diazo	negativ	☐	☐	☐	☐	☐
Sulkowitsch	Sulkow	negativ	☐	☐	☐	☐	☐
Porphyrin	Prophyrin	negativ	☐	☐	☐	☐	☐
Melanin	Melanin	negativ	☐	☐	☐	☐	☐
Indikan	Ind.	negativ	☐	☐	☐	☐	☐
Millon	Millon	negativ	☐	☐	☐	☐	☐
Bence Jones	Be. Jo.	negativ	☐	☐	☐	☐	☐
Phenolsulphthaleinpr.	PSP %		☐	☐	☐	☐	☐
PAH Clearance	PAH Cl.		☐	☐	☐	☐	☐
Inulin Clearance	In.-Cl.		☐	☐	☐	☐	☐
Endog. Kreat.-Clear.	K.-Cl. ml/min.		☐	☐	☐	☐	☐
a-Amylase	a-Am. mU	–350 mU	☐	☐	☐	☐	☐
Phenolkörper			☐	☐	☐	☐	☐
Osmolalität	mosm/l	800-900	☐	☐	☐	☐	☐
Spez.-Gewicht			☐	☐	☐	☐	☐
			☐	☐	☐	☐	☐

Resistenzbestimmung

+ = empfindlich (+) = schwach empfindlich Ø = resistent

Fig. 3 (contd.)

(will be continued)

Blutwerte

Anforderung ☒ — Istwerte vom:

Kategorie	Parameter	Referenz	Anf.	1	2	3	4
	Blutgruppe + RH-Faktor						☐
Blutbild	Hämoglobin	M: 14 - 18, F: 12 - 16, g %	☐	☐	☐	☐	☐
	Hämotokrit	36 - 44 %	☐	☐	☐	☐	☐
	Erythrozyten	4,5 - 5,5 Mill./mm³	☐	☐	☐	☐	☐
	Leukozyten	4000 - 9000 mm³	☐	☐	☐	☐	☐
	Diff. Blutbild — Basophile	0 - 1 %	☐	☐	☐	☐	☐
	Eosinophile	1 - 4 %	☐	☐	☐	☐	☐
	Stabkernige	1 - 5 %	☐	☐	☐	☐	☐
	Jugendliche		☐	☐	☐	☐	☐
	Segmentkernige	50 - 70 %	☐	☐	☐	☐	☐
	Lymphozyten	20 - 40 %	☐	☐	☐	☐	☐
	Monozyten	2 - 6 %	☐	☐	☐	☐	☐
	Sonstige		☐	☐	☐	☐	☐
Gerinnungswerte	Blutsenkung	1. Std.: 7 - 11, 2. Std.: 8 - 16 mm	☐	☐	☐	☐	☐
	Blutungszeit nach Duke	3 - 5 Min.	☐	☐	☐	☐	☐
	Gerinnungszeit	2 - 5 Min.	☐	☐	☐	☐	☐
	Antithrombin III	pathol. Bereich <70%	☐	☐	☐	☐	☐
	Thrombinzeit	16 - 20 Sec.	☐	☐	☐	☐	☐
	PTT	30 - 40 Sec.	☐	☐	☐	☐	☐
	Hepato-Quick-Test	70 - 130 % (therap. Bereich 10-20%)	☐	☐	☐	☐	☐
	Fibrinogen	150 - 450 mg %	☐	☐	☐	☐	☐
	Thrombozyten	150000-340000 mm³	☐	☐	☐	☐	☐
Herz	LDH	120 - 240 mU	☐	☐	☐	☐	☐
	Alpha-HBDH	55 - 140 mU	☐	☐	☐	☐	☐
	CK (NAC)	M: 10 - 80 mU, F: 10 - 70 mU	☐	☐	☐	☐	☐
	SGOT	M: – 18 mU, F: – 15 mU	☐	☐	☐	☐	☐
	SGPT	M: – 22 mU, F: – 17 mU	☐	☐	☐	☐	☐
	Gamma GT	M: 6 - 28 mU F: 4 - 18 mU	☐	☐	☐	☐	☐
Leber	Cholinesterase	3000 - 9300 mU	☐	☐	☐	☐	☐
	GLDH	M: – 4 mU, F: – 3 mU	☐	☐	☐	☐	☐
	LAP	11 - 35 mU	☐	☐	☐	☐	☐
	Ges. Eiweiß	6,6 - 8,7 g%	☐	☐	☐	☐	☐
	Dir. Bilirubin	– 0,25 mg %	☐	☐	☐	☐	☐
	ges. Bilirubin	–1,0 mg%	☐	☐	☐	☐	☐
	a·Amylase	–100 mU	☐	☐	☐	☐	☐
	Lipase	– 200 mU	☐	☐	☐	☐	☐
	Alk. Phosphatase	E: 60 - 170 mU, Kinder: 151 - 471 mU	☐	☐	☐	☐	☐
	Saure Phosphatase	– 11mU	☐	☐	☐	☐	☐
	Prostataphosphatase	– 4 mU	☐	☐	☐	☐	☐
Fette	Cholesterin	– 250 mg %	☐	☐	☐	☐	☐
			☐	☐	☐	☐	☐
	Neutralfett .	74 - 172 mg %	☐	☐	☐	☐	☐
Niere	Creatinin	M: 0,6 - 1,1, F: 0,5 - 0,9 mg %	☐	☐	☐	☐	☐
	Harnstoff	– 50 mg %	☐	☐	☐	☐	☐
	Harnsäure	M: 3,4 - 7,0 mg %, F: 2,4 - 5,7 mg %	☐	☐	☐	☐	☐
	Osmolalität	290 - 300 mOsm	☐	☐	☐	☐	☐
Elektrolyte	Natrium	132 - 150 mval/l	☐	☐	☐	☐	☐
	Kalium	3,8 - 5,0 mval/l	☐	☐	☐	☐	☐
	Calcium	4,4 - 5,5 mval/l	☐	☐	☐	☐	☐
	Chlorid	95 - 110 mval/l	☐	☐	☐	☐	☐
	Magnesium	1,7 - 2,6 mg %.	☐	☐	☐	☐	☐
	Phosphor anorg.	E: 2,5-5,0, Kinder 4,0-7,0 mg %	☐	☐	☐	☐	☐
Spurenel.	Eisen	M: 80 - 150 µ %, F: 60 - 140 µg %	☐	☐	☐	☐	☐
	Kupfer	65 - 165 µg %	☐	☐	☐	☐	☐
	Lithium	– 0,003 mval/l (therap. Bereich 0,5-1,5 mval/l)	☐	☐	☐	☐	☐
	Blutgasanalyse	pH 7,35 - 7,45	☐	☐	☐	☐	☐
		BE 0 ± 2	☐	☐	☐	☐	☐
		pCO₂ 35 - 45 mm Hg	☐	☐	☐	☐	☐
		pO₂ 80 - 90 mm Hg	☐	☐	☐	☐	☐
		Aktuelles Bic. 22 - 26 mEq	☐	☐	☐	☐	☐
Rheumat.	RF	<60 iu/ml	☐	☐	☐	☐	☐
	ASL		☐	☐	☐	☐	☐
	CRP	– 0,8 mg/dl	☐	☐	☐	☐	☐
	Blutzucker	Nüchtern: 70 - 100 mg % 7 h 30	☐	☐	☐	☐	☐
		10 h 30	☐	☐	☐	☐	☐
		13 h 30	☐	☐	☐	☐	☐
		16 h 30	☐	☐	☐	☐	☐
	COP (koll. osmot. Druck)	24 - 28 mm Hg					

Fig. 3 (contd.)

(will be continued)

206

Lungen-, Kreislauf- und Sonderdaten

Befund vom:

Anforderung: ☒		Normalwerte		
Ruhe EKG	☐	altersentsprechend unauf-fälliger Befund ☐		
Rö.-Thorax	☐	altersentsprechend unauf-fälliger Befund ☐		
Rö.-	☐	altersentsprechend unauf-fälliger Befund ☐		
Belastung EKG mit Watt	☐	altersentsprechend unauf-fälliger Befund ☐		
Kreislauffunktions-Test	☐	altersentsprechend unauf-fälliger Befund ☐		
Herzschall	☐			
Carotispuls	☐			
Lungenfunktion	☐			
Vitalogramm	☐			
Schilddrüse	☐			
Elektrophorese	☐			
Komplement C 3	☐	83 - 177 mg%		
Komplement C 4	☐	12 - 32 mg%		
Transferrin	☐	204 - 360 mg%		
Haptoglobin	☐	27 - 139 mg%		
Gamma - A Immunglobulin	☐	70 - 312 mg%		
Gamma - G-Immuglubulin	☐	639 - 1349 mg%		
Gamma - M-Immunglobulin	☐	56 - 352 mg%		
Phenolkörper i. Serum	☐			
Ammoniakspiegel i. S.	☐			
Poryhyrin-Körper	☐			
Australia-Antigen	☐			
Lactat	☐	0,63 - 2,44 mmol/l		
Sonstiges	☐			
	☐			

Fig. 3 (contd.)

A N A E S T H E S I O L O G I E

Sportkrankenhaus Hellersen

Chefarzt: Dr. R. Droh

Anaesthesie-Protokoll

	Name des Stationsarztes		Telefon-Nr. des Patienten	Zimmer-Nr. d. Patienten

Kartenart	Klinik	Nr.	Jahrgang	Geschl.	Tag	Geburtstag Monat	Jahr	Name	Mehr-ling
1	2 3	4 5 6 7 8 9 10 11		12 13	14 15 16	17	18 19	20	21

PRAEANAESTHETISCHER BEFUND

Besonderheiten

Station d. Pat. **22** — 1 amb. / 2 stat.

Adressette (bzw. Name [bei Frauen auch Mädchenname], Vorname, Geb.-Datum, Adresse)

geplante Operation:

Blutdruck / mmHg

Pulsfrequenz / min.

Größe | 23 24 25 | Gewicht | 26 27 28 29 | Typ
cm | kg
0 normal / 1 dünn / 2 muskulös / 3 dick

Zahnstatus
X Zahn fehlt
O Zahn locker

re | 7 6 5 4 3 2 1 | 1 2 3 4 5 6 7 | li.
7 6 5 4 3 2 1 | 1 2 3 4 5 6 7

30 Kreislauf
0 ohne path. Befund ☐ | 1 Hypotonie ‹ 100 syst. ☐ | 2 Volumendefizit ☐ | 3 Schock ☐ | 4 Hypertonie 160 syst. ☐ | 5 Pulsdefizit ☐
6 Ödeme ☐ | 7 Anämie ☐ | 8 ☐ | 9 keine Untersuchung ☐

31 Herz
0 ohne path. Befund ☐ | 1 Bradykardie ☐ | 2 Tachykardie ☐ | 3 Extrasystolie ☐ | 4 Arrhythmie ☐ | 5 kongen. Mißbildung ☐
6 erworbenes Vitium ☐ | 7 Zust. n. Infarkt ☐ | 8 ☐ | 9 keine Untersuchung ☐

Befunde:
EKG:
Rö.:

32 Atmungs-organe
0 o. path. Befund ☐ | 1 Asthma ☐ | 2 inspir. Stridor ☐ | 3 Emphysem ☐ | 4 Ruhedyspnoe ☐ | 5 Tbc aktiv ☐ | 6 Tbc alt ☐ | 7 Cyanose ☐
8 ☐ | 9 keine Untersuchung ☐

Befunde:
Rö.:

33 Urin-ausscheidung
0 o. path. Befund ☐ | 1 Albuminurie ☐ | 2 Haematurie ☐ | 3 Oligurie/Anurie ☐ | 4 Polyurie ☐ | 5 Glucosurie ☐ | 6 Acetonurie ☐ | 7 Bakteriurie ☐
8 ☐ | 9 keine Untersuchung ☐

34 Stoffwechsel
0 o. path. Befund ☐ | 1 Fieber ☐ | 2 Diabetes ☐ | 3 Hyperthyreose ☐ | 4 Uraemie ☐ | 5 Hypokaliaemie ☐ | 6 Acidose ☐ | 7 Alkalose ☐
8 ☐ | 9 keine Untersuchung ☐

Befunde:
Na. | K. | Ca. | Cl. | OT. | PT. | y-GT. | HST. | Creatinin | BZ
Ges. Eiweiß | Klinische Befunde

35 Dauer-medikation
0 keine ☐ | 1 Insulin ☐ | 2 Cortison ☐ | 3 Cardiaca ☐ | 4 Antihypertonica ☐ | 5 Analgetica ☐ | 6 Sedativa ☐ | 7 Schilddrüsenprä. ☐
8 ☐ | 9 keine Untersuchung ☐

36 Allergie
0 keine ☐ | 1 Pflaster ☐ | 2 Jod ☐ | 3 Antibiotika ☐ | 4 Anaesthetika ☐ | 5 Analgetica ☐ | 7 Unbek. Allergien ☐
8 ☐ | 9 keine Aussage ☐

37 Weitere Befunde (frühere Op.)
0 keine | 1 Hepatitis floride | 2 Hepatitis früher | 3 Fettleber | 4 Zirrhose | 8

38 Letzte Nahrungsaufnahme
1 › 12 Std. | 2 6-12 Std. | 3 4-6.Std. | 4 ‹ 4 Std. | 9 keine Aussage

39 Risikogruppe
I | II | III | IV | V | VI | VII

PRAEMEDIKATION

Bericht an den Hausarzt angefertigt am:

Vorabend | Kaps. Limabatril | Tab. Rohypnol

☐ Hausmedikation exakt weiterführen
☐ Ery-Konzentrat bestellen — ml
☐ Frischblut bestellen — ml
☐ Mineraldrinks am Tag vor OP — Liter
☐ Heparin – Dihydergot täglich s. c. — 1 Amp.
Erstinjektion jeweils zeitgleich mit Prämedikationsspritze
☐ Liquemin täglich s. c. — 5000 IE
Erstinjektion jeweils zeitgleich mit Prämedikationsspritze.
☐ Rohypnol auf Abruf vor OP — mg. i. m.
☐ wegen Hämodilution täglich Eryfer — 3 x 1 Kps.
☐ Kalinor — Drg.

Dat. der Prämed. | Prämed. Arzt

Uhrzeit | Unterschrift der Schwester | 41 | 42 | * Narkose hier

POSTOP.

43 Kompl. nach 36 Stunden
0 keine ☐ | 1 metabolisch ☐ | 2 Exitus ☐ | 3 pulmonal ☐ | 4 respiratorisch ☐ | 5 kardial ☐ | 6 zirkulatorisch ☐ | 7 renal ☐
8 ☐ | amb. Patient ☐

44 Sektion ☐ nein ☐ ja | Befund

Rechte Spalte (Codierung):
1 amb. / 2 stat. — 22
Größe — 23 24 25
Gewicht — 26 27 28
Typ — 29
Kreislauf — 30
Herz — 31
Atmungsorgane — 32
Urinausscheidung — 33
Stoffwechsel — 34
Dauermed. — 35
Allergie — 36
Weitere Befunde — 37
Letzte N. A. — 38
Risikogruppe — 39
Praemedikation — 40
41 42
43
44

Fig. 4

(will be continued)

ANAESTHESIEVERLAUF

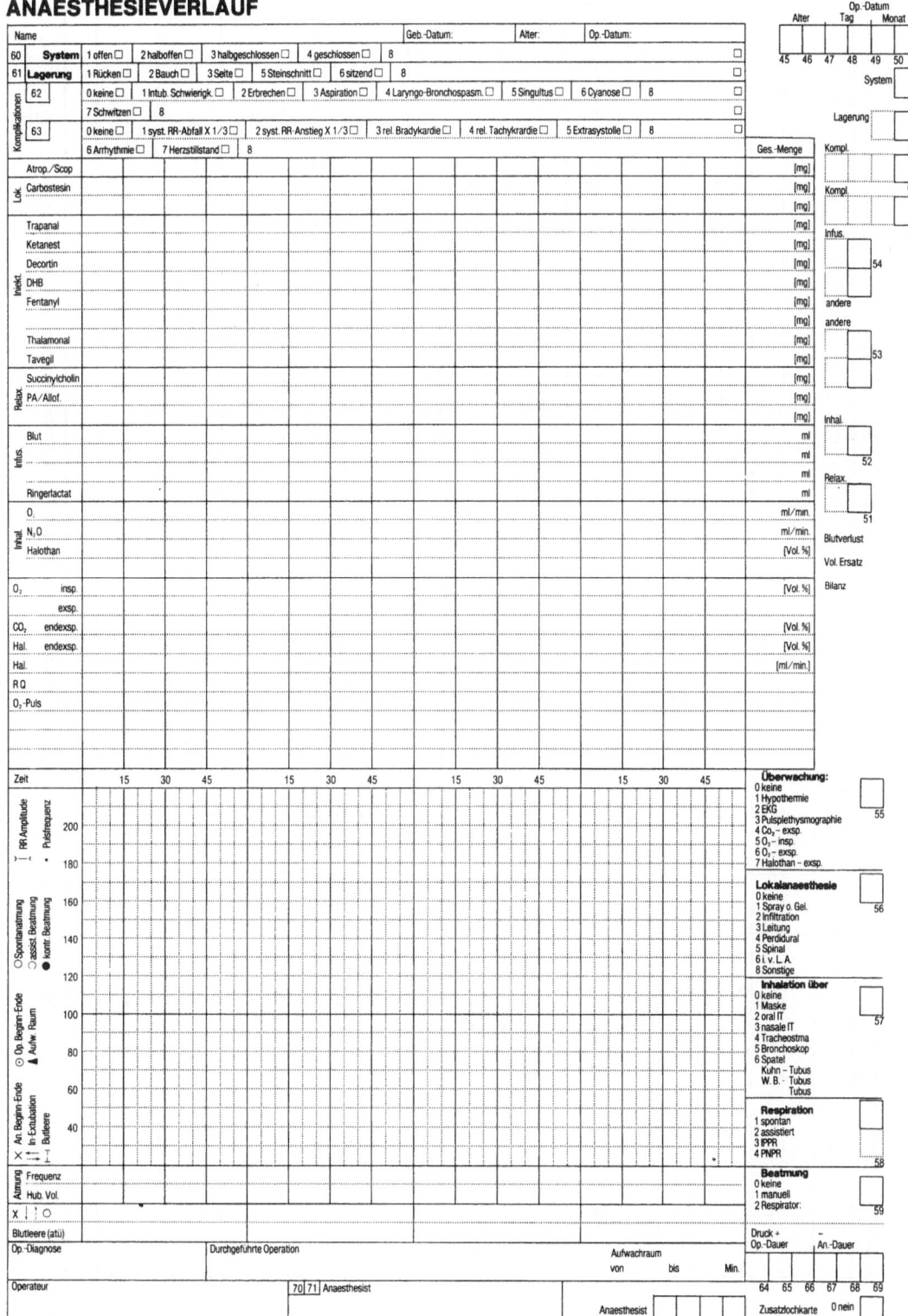

Fig. 4 (contd.)

Anaesthesiologische Besonderheiten aus der Praemedikationsvorstellung vom

Name / Station	Spinal	Plexus	Peridural	Diabetes	Kleinkind	Alte TBC	1 Amp. Decadron	Arzt zur Einleitung	Alupent 1 : 9	Hämo-dilution	Plasma Blut	16er Kanüle+RL	E – lyte – Subtitution	Sonstiges

(Patienten, die noch nicht auf dem OP-Plan erschienen sind, bitte auf das nachfolgende Blatt übertragen)

Fig. 5

ANAESTHESIOLOGIE

Sportkrankenhaus Hellersen
Chefarzt: Dr. R. Droh

An.-Vorbereitungs-Kontrolliste
– Allgemein - Anaesthesie –

(bei Frauen auch Mädchenname)

Name:

Vorname:

Geb.-Datum:

1) Haben Sie den Patienten nach seinem Namen gefragt
und stimmt dieser mit dem auf dem Op.-Plan
und Anaesthesieprotokoll überein ? . . . . ☐ Ja ☐ Nein

2) Stimmt nach Angaben des Patienten die angezeichnete Lokalisation seiner
bevorstehenden Operation mit dem Op.-Plan überein ? . . . . ☐ Ja ☐ Nein

3) a) Ist der Patient nüchtern ? . . . . ☐ Ja ☐ Nein
b) seit wie vielen Stunden ? _____

4) Ist ein evtl. vorhandenes Gebiß des Patienten entfernt ? . . . . ☐ Ja ☐ Nein

5) Sind die Zähne des Patienten fest ? . . . . ☐ Ja ☐ Nein

6) Ist die Mundhöhle ohne Besonderheiten ? . . . . ☐ Ja ☐ Nein

7) Ist das Intubationsset einschließlich sämtlich aufgezogener Spritzen vollständig ? . . ☐ Ja ☐ Nein
8) Ist die Manschette des Tubus dicht ? . . . . ☐ Ja ☐ Nein

9) Wurde das Tubuskaliber dem Geschlecht sowie Alter angepaßt
und der Woodbridge-Tubus mit einem gleitenden Führungsstab versehen ? . . . . ☐ Ja ☐ Nein

10) Wurde der Tubus gleitfähig gemacht ? . . . . ☐ Ja ☐ Nein

11) Ist die Beleuchtung des Laryngoskops in Ordnung ? . . . . ☐ Ja ☐ Nein

12) Erfolgte die Kontrolle von Puls und Blutdruck ? . . . . ☐ Ja ☐ Nein

13) Ist das Narkosesystem dicht ? . . . . ☐ Ja ☐ Nein
14) Wurden Krankenfragebogen, Anaesthesieprotokoll
und Laborbogen auf Ihre Vollständigkeit kontrolliert ? . . . . ☐ Ja ☐ Nein

15) Liegt der Anaesthesieaufklärungsbogen unterschrieben vor ? . . . . . . ☐ Ja ☐ Nein

16) Hat der Patient in der Nacht gut geschlafen ? . . . . ☐ Ja ☐ Nein

17) Sind Gewicht, Größe und die Laborwerte im Narkoseprotokoll vermerkt ? . . . . ☐ Ja ☐ Nein

18) Wurde eine mit Nein angekreuzte Frage dem Anaesthesiearzt gemeldet ? . . . . ☐ Ja ☐ Nein

19) Liegt die Operations- und Anaesthesieerlaubnis der Erziehungsberechtigten vor ? . . ☐ Ja ☐ Nein

20) Liegt ein steriler Absaugkatheter bereit ? . . . . ☐ Ja ☐ Nein

21) Hat der Patient seine Praemedikation erhalten ? . . . . ☐ Ja ☐ Nein

22) Wurde das EKG-Kissen untergelegt ? . . . . ☐ Ja ☐ Nein

23) Wurde der Puls-Fingerling angelegt ? . . . . ☐ Ja ☐ Nein
24) Wurde der Kopfhörer aufgesetzt ? . . . . ☐ Ja ☐ Nein

An.-Datum   Unterschrift

Fig. 6

(will be continued)

– Peridural- bzw. Spinalanaesthesie –

1. Wurde der Patient 60min. vor dem voraussichtlichen Op.-Beginn
    in den Vorbereitungsraum bestellt? . . . . □ Ja   □ Nein

2. Ist der Kopf-Rückenteil des vorgesehenen Op.-Tisches bis auf 35° hochkurbelbar? . . . □ Ja   □ Nein

3. Ist der Patient nüchtern? . . . . □ Ja   □ Nein

4. Wurden dem Patienten Haube und Mundschutz angelegt? . . . . □ Ja   □ Nein

5. Liegen folgende Patientenpapiere vor:
    a) die schriftliche Aufklärung des Patienten? . . . . □ Ja   □ Nein
    b) der Anaesthesie-Fragebogen? . . . . □ Ja   □ Nein
    c) das Anaesthesie-Protokoll? . . . . □ Ja   □ Nein
    d) die Einwilligung der Eltern bei Minderjährigen? . . . . □ Ja   □ Nein

6. Wurde folgendes gerichtet bzw. im Regionalanaesthesiekasten bereitgestellt:
    a) vollständiges Intubationsset mit Trapanal- und Succinylampullen
        sowie entsprechenden Spritzen unaufgezogen? . . . . □ Ja   □ Nein
    b) Blutdruckmeßapparat und Stethoskop? . . . . □ Ja   □ Nein
    c) EKG-Kissen? . . . . □ Ja   □ Nein
    d) Dermograph? . . . . □ Ja   □ Nein
    e) Einmalrasierer? . . . . □ Ja   □ Nein
    f) sterile Handschuhe Größe 7,5 und 8? . . . . □ Ja   □ Nein
    g) Alkohol zur Entfettung der Haut? . . . . □ Ja   □ Nein
    h) Spray zur Hautdesinfektion? . . . . □ Ja   □ Nein
    i) Set der Firma Sherwood Monojekt
        1. Spinal 22 gauche, 2. Spinal 25 gauche, 3. Peridural und zusätzliches, steriles Millipore-Filter? □ Ja   □ Nein
    j) 0,5%iges Carbostesin ohne Adrenalin, 4 Amp. à 5 ml? . . . . □ Ja   □ Nein
    k) eine Amp. 0,9%iges NaCl, 10 ml? . . . . □ Ja   □ Nein
    l) eine 10 ml Spritze mit Effortil 1 : 9? . . . . □ Ja   □ Nein
    m) eine 2 ml Spritze mit 0,5 mg Atropin? . . . . □ Ja   □ Nein
    n) eine Ampulle Valium (10 mg) und eine 2 ml Spritze unaufgezogen? . . . □ Ja   □ Nein
    o) Äther zur Sensibilitätstestung? . . . . □ Ja   □ Nein
    p) Tupfer und Pflaster? . . . . □ Ja   □ Nein
    r) Kanülen Nr. 2 und Butterfly Kanülen? . . . . □ Ja   □ Nein

7. Instrumentiertischchen? . . . . □ Ja   □ Nein

8. Einsatzbereites Narkosegerät? . . . . □ Ja   □ Nein

9. Wurde eine Ringer-Lactat-Infusion angelegt? . . . . □ Ja   □ Nein

10. Sitzt der Patient auf dem horizontalen Teil des Op.-Tisches
    und stützt seine Beine auf einem Stehhocker ab? . . . . □ Ja   □ Nein

11. Wurde nach Abschluß der Vorbereitungen der Anaesthesiearzt verständigt? . . . . . . □ Ja   □ Nein

Datum: ....

Uhrzeit: ....

Unterschrift der Anaesthesieschwester bzw. Anäesthesiepfleger

Fig. 6 (contd.)

A N A E S T H E S I O L O G I E

Sportkrankenhaus Hellersen
Chefarzt : Dr. R. Droh

01) Emma
Have you: – prewarmed the apparatus at least 0.5 h before commencement
of anaesthesia?   Yes □ No □
– set the indicator to 0 before first narcosis?   Yes □ No □
– tested the indicator at the preset control value
before narcosis?   Yes □ No □
– checked that the transducer fits the apparatus?   Yes □ No □
– put the transducer in the right direction of current
in the expiration part of the apparatus?   Yes □ No □

02) Datex CD 101 CO_2 without O_2
Have you: – turned on at least 0.5 h before commencement of anaesthesia? Yes □ No □
– correctly connected the inflow and outflow tubes to the
anaestetic circuit, attached the outlet to "low flow"?   Yes □ No □
– set the indicator to O before commencement of anaesthesia?   Yes □ No □

03) Datex Normocap CO_2 with O_2
Have you: – turned on and set the apparatus as for Datex CD 101?   Yes □ No □
– also set O_2 to 21% (small screw labelled O_2 on back of
apparatus)?   Yes □ No □

04) ECG and Pulse
Have you: – turned on ECG machine and pulse recorder 0.5 h before
commencement of anaesthesia?   Yes □ No □
– checked that ECG and pulse leads are perfectly in order?   Yes □ No □
– checked that the ECG pads and electrodes are properly
connectly?   Yes □ No □

05) Anaesthetic circuit
Have you: – checked that all parts are properly attached?   Yes □ No □
– checked that O_2 and N_2O tubes have not been confused?   Yes □ No □
– checked whether gas pressure stays constant when the
anaesthetic circuit is filled with O_2 and sealed off?   Yes □ No □
– checked that all ventilator flaps move freely?   Yes □ No □
– checked that all connectors are properly joined to one another
and that all points of connection (metal connectors) have
been lubricated before assembly of the apparatus?   Yes □ No □

06) Oxycom
Have you: – checked the battery voltage?   Yes □ No □
– checked the alarm limit?   Yes □ No □
– calibrated to 21% or 80%–100%, as appropriate?   Yes □ No □
– paid attention to the sintching from setting to setting?   Yes □ No □

07) Halothane Vaporizer
Have you: – filled the Vaporizer with halothane up to the maximal level Yes □ No □
– checked that the inlet and outlet are correctly attached?   Yes □ No □

08) Aspiration Unit
Have you: – completely assembled the unit?   Yes □ No □
– filled a container with water?   Yes □ No □
– checked that the unit functions properly?   Yes □ No □

(will be continued)

Fig. 7. Equipment checklist before first narcosis. (Translated from the German)

09) Sphygmomanometer
Have you: – checked that the sphygmomanometer is airtight?   Yes □ No □
           – checked that the pressure indicator works?   Yes □ No □
           – checked that there is a stethoscope?   Yes □ No □

10) Music System
Have you: – checked that you have three good pairs of headphones?   Yes □ No □
           – checked that you have an extension cable?   Yes □ No □
           – checked the loudspeaker?   Yes □ No □
           – checked that the music reproduction is perfect?   Yes □ No □

Date: ....   Signature: ....

Fig. 7 (contd.)

214

A N A E S T H E S I O L O G I E

Sportkrankenhaus Hellersen
Chefarzt : Dr. R. Droh

Überwachungsbogen des Aufwachraumes	Aufwachraumbuch

Der Überwachungsbogen legt fest:

1. wie oft der *Blutdruck* zu messen und einzutragen ist

2. wie oft der Puls zu messen und einzutragen ist

3. ob und wie oft die *Temperatur* zu messen ist

4. ob und wie oft die Atmung zu messen ist

5. ob, mit welchem *Volumen* und welcher *Atemfrequenz* zu beatmen ist

6. ob, und wie oft *Blutverluste* aus Dränagen zu messen sind

7. ob, wie oft und welche *Sekrete* zu messen sind

8. ob und wie oft der zentrale Venendruck zu messen ist

9. besondere Anweisungen müssen zusätzlich klar geschrieben und formuliert festgelegt oder verneint werden

10. ob, welche *Infusionen* und wieviel in welchem Zeitraum einlaufen soll

11. ob, welche und wieviel in welchem Zeitraum an *Blut*, Blutfraktionen oder Blutersatzmittel transfundiert bzw. infundiert werden sollen

12. ob, welche und wieviel *Arzneimittel* in welchen Zeiträumen bzw. zu welchen Zeitpunkten s.c., i.m., i.v. oder i.a. verabreicht werden sollen

13. *Besondere Anordnungen* müssen auch hier entweder klar und deutlich verneint oder/ festgelegt werden

14. ob, wann und welche *Laborwerte* bestimmt werden müssen, z. B. HB, Ery., HK, Blutgasanalysen, BZ, Elektrolyte, Pulskurven, EKG, Gerinnungsstatus, Gesamteiweiß, Elektrophorese, usw.

Zu empfehlen sind die folgenden Aufzeichnungen:

1. Name und Vorname

2. Alter

3. Fortlaufende Anaesthesienummer

4. Station

5. Datum

6. Uhrzeit-Zugang

7. Uhrzeit-Abgang

8. Narkoseform (Allgemeinanaesthesie/Lokalanaesthesie)

9. Spezielle Anaesthesieformen

10. Besondere Maßnahmen wie Absaugen, O_2-Verabreichung etc.

11. Postoperative Beatmung

12. Komplikationen

13. Puls

14. Blutdruck

15. Kopfanheben

16. Hautkolorit

17. Atmung regelmäßig und unbehindert

18. Ansprechbarkeit

19. Zeitlich und örtlich orientiert

20. Schmerzmittel

21. Unterschrift der verantwortlichen Pflegekraft.

Fig. 8

215

A N A E S T H E S I O L O G I E

Sportkrankenhaus Hellersen

Chefarzt Dr. R. Droh

Postanaesthesie-Fragebogen

Prämedikation abends ....

Prämedikation morgens ....

Halothan ....

N20 / 02 ....

Fentanyl / Thalamonal ....

DHB ....

Ketanest ....

Extremitäten-Chirugie ....

Bauch-Chirugie ....

Lagerung :   R: B:

Nicht vom Patienten auszufüllen !

Adressette (bzw. Name [bei Frauen auch Mädchenname], Vorname, Geb.-Datum, Adresse)

Sehr verehrte Patientin!

Sehr geehrter Patient!

Bitte füllen Sie diesen Fragebogen am Tage nach der Narkose aus. Wir erhoffen uns von Ihnen Angaben evtl. Hinweise zur weiteren Verbesserungsmöglichkeit unserer Technik und Arbeit für Sie und die Patienten nach Ihnen. Beantworten Sie bitte **jede** Frage mit „ja" oder „nein" mit einem Kreuz bzw. mit einer handschriftlichen Bemerkung, da wir unvollständig ausgefüllte Fragebögen leider nicht verwerten können. Besten Dank für Ihre Mühe.

Fragen:

Alter:...............Geschlecht:...............Größe:...............Gewicht:...............Beruf:...............

A) Hatten Sie Angst vor der Narkose ? . . . .   ○ ja   ○ nein

Wenn ja, warum ? ....

B) Hätten Sie vor der Narkose gerne noch weitere Erklärungen gehabt ? . . . .   ○ ja   ○ nein

Wenn ja, welche ?

C) Wie haben Sie in der Nacht vor der Narkose geschlafen ?

1) schlecht   . . . .   ○ ja   ○ nein

2) gut   . . . .   ○ ja   ○ nein

3) sehr gut   . . . .   ○ ja   ○ nein

D) Welche der folgenden Möglichkeiten entsprach am ehesten Ihrem Zustand, nachdem Sie auf der Station Ihre Beruhigungsspritze bekommen hatten ?

1) Keine Müdigkeit ?   . . . .   ○ ja   ○ nein

2) Zunehmende Müdigkeit ?   . . . .   ○ ja   ○ nein

3) Fortdauernde Unruhe ?   . . . .   ○ ja   ○ nein

4) Schläfrigkeit und Gleichgültigkeit ?   . . . .   ○ ja   ○ nein

5) Vor der Spritze relative Ausgeglichenheit, danach zunehmende Aufgeregtheit und unmotivierte heftige Angstgefühle ?   . . . . .   ○ ja   ○ nein

E) Kam nach der Vorbereitungsspritze auf der Station :

Ein Übelkeitsgefühl auf ?   . . . .   ○ ja   ○ nein

Wenn ja, mußten Sie erbrechen ?   . . . .   ○ ja   ○ nein

F) 1) Welche Art von Musik hatten Sie sich vor der Narkose gewünscht ?

....

2) Welche Art von Musik hörten Sie vor der Narkose ?

....

3) Was hat Ihnen an der gehörten Musik gefallen, und was hat Ihnen daran nicht gefallen ?

....

4) Nachdem Sie nun (die von Ihnen gewünschte) Musik gehört haben, würden Sie sich für eine evtl. erneute Narkose die gleiche Musik wünschen ?   . . . .   ○ ja   ○ nein

Wenn nein, warum nicht ? ....

Welche andere Musik würden Sie stattdessen bevorzugen ?

....

Warum ? ....

5) **Bevorzugten Sie vor der Narkose absolute Ruhe ?**   . . . . .   ○ ja   ○ nein

Wenn ja, warum ? ....

Fig. 9   *(will be continued)*

216

6) Nehmen Sie regelmäßig eines der folgenden Medikamente ein ?
   a) Schlafmittel . . . . O ja   O nein
   a) Wenn ja, welches ? ....
   b) Beruhigungsmittel ? . . . . O ja   O nein
   b) Wenn ja, welches ? ....
   c) Psychopharmaken . . . . O ja   O nein
   c) Wenn ja, welches ? ....
   d) Schmerzmittel . . . . O ja   O nein
   d) Wenn ja, welches ? ....

G) Bemerkten Sie das Einschlafen zur Narkose ? . . . . O ja   O nein
   Wenn ja: ....
   1) Haben Sie dabei etwas besonderes gehört ? . . . . . O ja   O nein
   Wenn ja, was ? ....
   2) Haben Sie dabei etwas besonderes gesehen ? . . . . . O ja   O nein
   Wenn ja, was ? ....
   3) Hatten Sie dabei ein besonderes seelisches Erlebnis (Angst, Hochgefühl oder dergleichen) ? O ja   O nein
   Wenn ja, bitte näher beschreiben ? ....
   4) Ließen Sie die Narkoseerlebnisse gleichgültig ? . . . . . O ja   O nein

H) Hatten Sie Träume ? . . . . O ja   O nein
   Wenn ja, bitte näher beschreiben ? ....

I) Hatten Sie sonst irgendwelche Wahrnehmungen ? . . . . . O ja   O nein
   Wenn ja, bitte näher beschreiben ? ....

J) Erinnern Sie sich, wann Sie nach der Narkose wieder wach waren ? . . . O ja   O nein
   Etwa Uhrzeit : ....

K) Wann trat der erste Wundschmerz auf (etwa Uhrzeit) ? ....

L) Hatten Sie nach dem Erwachen aus der Narkose
   1) Schwindel ? . . . . O ja   O nein
   2) Übelkeit ? . . . . O ja   O nein
   3) Erbrechen ? . . . . O ja   O nein
   4) Schluckbeschwerden ? . . . . O ja   O nein
   5) Halsbeschwerden ? . . . . O ja   O nein
   6) trockenen Mund ? . . . . O ja   O nein
   7) Katergefühl ? . . . . O ja   O nein
   8) Gliederschmerzen ? . . . . O ja   O nein
   9) Schwächegefühl ? . . . . O ja   O nein
   10) Sprachstörungen ? . . . . O ja   O nein
   11) Sehstörungen ? . . . . O ja   O nein
   12) Kopfschmerzen ? . . . . O ja   O nein
   13) Muskelkater ? . . . . O ja   O nein
   Bitte beschreiben Sie, wie Sie sich nach der Narkose gefühlt haben, wenn von 1 - 13 nichts zutrifft. ....
   ....

M) Wie fanden Sie die Narkose ?
   1) angenehm . . . . O ja   O nein
   2) unangenehm . . . . O ja   O nein
   3) keine Meinung . . . . O ja   O nein
N) Würden Sie sich eine solche Narkose, **wenn nötg** nochmals wünschen ? . . . O ja   O nein
   Bemerkungen : ....
O) Hatten Sie früher schon mal eine Vollnarkose ? . . . . O ja   O nein
   Wenn ja, wie fanden Sie die jetzige Narkose im Gegensatz zur Vorigen ?
   1) besser . . . . O ja   O nein
   2) schlechter . . . . O ja   O nein
   3) gleich angenehm . . . . O ja   O nein
   Sonstige Bemerkungen: ....
P) Haben Sie Astronautenkost erhalten ? . . . . O ja   O nein
   Wenn ja, wie beurteilen Sie diese und die dazu erhaltene Aufklärung ? ....

Q) Zusätzliche Beobachtungen : ....
   ....
   ....
   ....

Datum : .... Unterschrift : ....

Fig. 9 (contd.)

Development and Practical Use of a Computerized Anaesthesia Protocol

L. Weller, H. J. Hartung, P. M. Osswald, H. J. Bender, and H. Lutz

The making of an anaesthesia record is to be seen as an absolute requirement, and there is a large amount of data and findings to be recorded. Information on the preoperative condition and the case history is essential for decisions on the intra- and postoperative course of the anaesthesia, and contributes decisively to the risk interpretation. Individually, this is personal data on the patient, especially details of serious previous and current illnesses, prescribed premedication, anaesthetic procedure, operation to be performed and names of anaesthetists and surgeons. The intraoperative data recording must take into account a variety of off- and on-line parameters, medicaments applied during the anaesthesia and all vital parameters of respiration and circulation; complications and any other special features are to be recorded and documented in a timely manner. Furthermore, fluid and volume balances must be made over the period of the anaesthesia, and venous and arterial approaches must be observed.

Conventional anaesthesia record making, however, allows only discontinuous documentation of all recorded and controlled parameters, so that important information is lost or can be subjectively influenced. In order to achieve a more comprehensive picture of the course of the whole anaesthesia, a computer system was developed at the Institute of Anaesthesiology and Reanimation in Mannheim which continually records and presents biosignals of haemodynamics, ventilation and administered ventilation gases via suitable sensors. Apart from this, various off-line parameters, such as infusions or medicaments given, can be documented in a timely manner.

Concept

The aim is the integration of monitors already available and in daily use into a flexible system of data recording and data presentation. The anaesthetic control system consists of a microprocessor – situated in the operating theatre – and monitors connected in series, through which the prior recording of the patient's parameters takes place via sensors [12, 16]. The ventilation parameters include intake volume, pressures, ventilation minute volume, expiratory resistance, dead space ventilation, compliance and expiratory CO_2 concentration. The concentrations of the ventilation gases administered – oxygen, nitrous oxide, halothane – are also continually recorded. Haemodynamic monitoring includes recording of heart rate and pulmonary and general blood pressures. Thus up to 16 on-line parameters can be processed and represented graphically. All on- and off-line parameters recorded can, at the end of the anaesthesia, be printed in the form of medical charts. Furthermore, all values measured can be printed out digitally in tabular form.

Generation

Recording and Presentation of Preoperative Data

Before commencement of the anaesthesia certain important patient data and organisatory information must be ascertained. This information is recorded with the help of a form which is set up on the screen (Fig. 1).

On-line Data Recording, Presentation

Up to 16 on-line parameters can be documented at 1-min intervals. The measured data is input via an analog-digital transformer, to which every monitor with a linear signal output between -10 V and $+10$ V can be connected. The data output appears either in graphic form, in so-called trends, or as columns of figures. In the graphic output, up to five parameters in two coordinate systems are simultaneously listed against time; the time axis can be selected from fixed intervals. Since the patient's circulation pattern during the anaesthesia can be regarded as basic information, systolic and diastolic blood pressure and heart rate must be represented in a graph, so that in each case two additional freely selected parameters

PREMEDICATION

NO. OF MEDICAMENTS ADMINISTERED (0–10): **3**

DATE	MEDICAMENT	DOSE (MG)	ROUTE	TIME
280682	ATROPIN	0.5	I.M.	10.15
	DOLANTIN	50		
	PSYQUIL	20		

WHERE ADMINISTERED	ADMINISTERED BY	EFFECT
HOSPITAL **Y**	SELF **Y**	SUFFICIENT **Y**
ELSEWHERE	OTHER	INSUFFICIENT

PREOPERATIVE STATUS

WEIGHT: **75**   HEIGHT (CM): **162**   STARRED (H): **6**

BLOOD PRESSURE: **110/70**   BLOOD GROUP: **A +**   DENTAL STATUS: **FIRM**

HR: **90**   HB: **11.1**   HCT: **33**   BS: **74**   K: **4.2**   NA: **145**

RISK GROUP: **3**

Fig. 1. Protocol for preoperative data

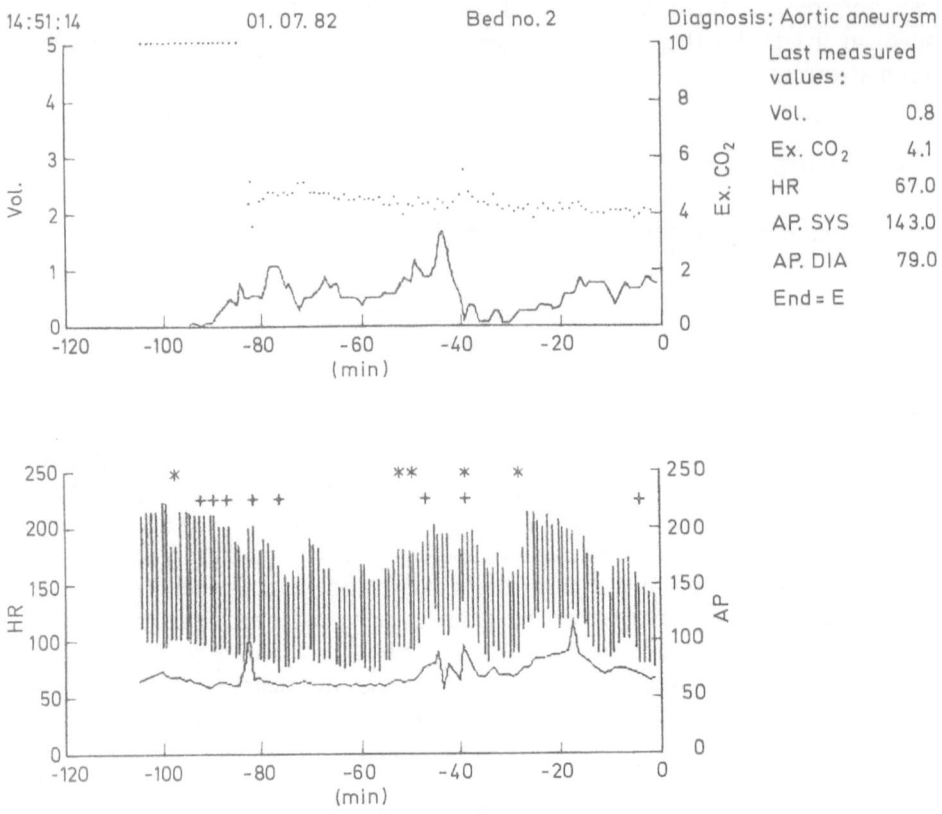

Fig. 2. On-line data featur heart rate and blood pressures as well as a choice of two other parameters

can be shown. In addition, the current measured values appear as large numbers on the right-hand side of the screen. On the top edge of the screen the current time and patient identification are displayed (Fig. 2).

Off-line parameters can be fed in and stored at 20-min intervals. From on- and off-line measured values the microprocessor calculates seven further parameters, which round off the picture of the patient's clinical condition. The balance of medication input and output is automatically computed, and time and form of administration, speciality and dose are documented. Furthermore, the circulation graph indicates at what specific times medication, infusion or transfusion was undertaken (Fig. 3).

Postoperative Details

The anaesthesia record concludes with the postoperative details, as well as a free-text input. Thus, with the aid of relevant forms, details of the anaesthetic process, the positioning of the patient, the area of operation, anaesthesia and operation times, possible intraoperative complications, the direct postoperative condition of the patient, and the location to which he has been transferred are all documented. Beyond this, it is possible to record free-text inputs, for example measures to be taken directly after the operation (Fig. 4).

220

LIST OF MEDICAMENTS

NO.	TIME	TYPE		PREPARATION	DOSE (ML, MG)
01	1031	K		RINGER	500
02	1035	M		FENTANYL	0.20
03	1045	M	ALLO		2
04	1045	M	VALIUM		10
05	1045	M	HYPNO		12
06	1045	M	SUCCI		80
07	1047	M	ALLO		6
08	1052	K		RINGER	500
09	1101	C		HAES	500
10	1108	K		RINGER	500
11	1130	C		HAES	500
12	1132	M	FENTANYL		0.15
13	1150	B		E2196267	250
14	1207	K		RINGER	500
15	1217	K		BICARBO.	200
16	1229	B		E2196267	250
17	1245	M	ALLO		2
18	1254	B		E2196319	250
19	1302	K		RINGER	500
20	1303	U		URINE	670
21	1304	D		BLOOD IN DRAIN	300
22	1332	C		HA 5%	400
23	1332	B		E2196246	250
24	1341	B		E2196326	250
25	1355	K		BAS	500

Fig. 3. Medication information includes time, type of administration, preparations used and dose. *K*, crystalloid; *C*, colloid; *M*, medicament; *B*, blood; *U*, urine; *D*, drain

COMPLICATIONS

FAILED VEIN PUNCTURE –	HAEMATOMA –	SEVERE HYPOTENSION –
DIFFICULT INTUBATION –	NERVE DAMAGE –	HYPERTONIC REACTION –
DENTAL PROBLEMS –	FAILED PUNCTURE –	BRADY-/TACHYCARDIA –
EQUIPMENT FAILURE –	INSUFF. ANAESTHESIA –	EXTRASYSTOLES –
ACCIDENTAL ARTERIAL PUNCTURE –		ASYSTOLE –

ALLERGIC REACTION –	LARYNGOSPASM –	VOMITING –
ANAPHYLACTIC REACTION –	BRONCHOSPASM –	SINGULTUS –
TRANSFUSION REACTION –	ASPIRATION –	MENTAL CHANGES –
TOXIC REACTION –	ATELECTIASIS –	POSITIONAL DAMAGE –
MALIGNANT HYPERTENSION –	PNEUMOTHORAX –	HEAT DAMAGE –
		SEVERE BLEEDING –

POSTOPERATIVE:	TRANSFERRED TO:
COOPERATIVE –	GENERAL WARD –
SOMNOLENT **Y**	OBSERVATION WARD –
SOPORIFIC –	INTENSIVE CARE –
	RECOVERY ROOM **Y**

POSTOP. MEASURES

HB, HCT, ACID-BASE (ASTRUP), BS, CHEST XR

Fig. 4. Protocol for postoperative data, e.g. complications and "free text"

Discussion

In the area of intensive care medicine the continual monitoring and documentation of parameters has become routine practice, brought about partly by the industry and partly by the microprocessor systems developed by working groups. Most predominantly, cardiocirculatory values are recorded online. For respiration, there is usually only the possibility of recording the frequency continually. Individual centres have worked out programs for the on-line recording of pulmonary parameters. In the area of anaesthesia, no suitable microprocessor system is offered commercially which meets the special requirements of this field for the measurement of cardiocirculatory parameters and pulmonary values.

Individual anaesthesiological institutes have themselves tried to find an adequate solution, with the idea of replacing the handwritten anaesthesia record as far as possible. Systems have been conceived in Sweden and the USA which guarantee online monitoring of important anaesthetic values. The emphasis of these systems, however, is on recording and documenting data; a suitable transparent data recording, which would make the records irrelevant, is not offered.

A further development can be seen in Atlanta, USA, unique so far in producing complete computer recording [2, 4, 6, 11, 13, 14, 18–24].

The system developed at the Institute of Anaesthesiology in Mannheim for anaesthesiological purposes has, for over 20 months now, proved successful in various operation fields, especially neurosurgery and routine vascular and abdominal surgery. It guarantees the recording of relevant on- and off-line parameters, and allows the continuous recording and presentation of the measured values.

The practical selection of parameters is of central significance. From the large amount of biosignals which it is today technically possible to record, the most informative for routine use on one hand and for scientific formulations and interests on the other must be selected, whereby no additional risk for the patient may arise [9]. Primarily these are the heart rate signals, which during the narcosis, together with the systemic pressures, represent the cornerstones of narcosis monitoring, since all anaesthetics influence these values [5].

The presentation of these parameters has a decisive significance, so that systemic pressure and heart rate are always displayed on the screen as basic information, and are measurable with non-invasive methods [1] (DINAMAP, Criticon Comp.). Further haemodynamic values must mainly be determined by invasive techniques, whereby continuous recording is also possible [7, 8].

Ventilatory measured values are obtained by non-invasive methods. As a measure of sufficient ventilatory minute volume, expiratory CO_2 is documented with intake volume. Blood gas determinations only allow point measurements of the $PaCO_2$, and in addition, can only be obtained reliably through arterial puncture. End-expiratory CO_2 concentration is a measured value that can be obtained simply. Oxygen concentration and narcotic gas concentrations are monitored continuously. In this way hypoxia through technical failure is identified. Changes in lung mechanics – through obstruction, insufficient relaxation, change in position of the tubus, shifting of the patient, etc. – are recorded and documented by changes in compliance or expiratory resistance, and appropriate therapeutic measures are taken without delay [15].

The application of the system is simple, since the monitors and monitoring equipment already available can be used, and the anaesthetist is familiar with these machines. Therefore the staff administering the anaesthesia are not distracted by the technology, but effectively supported by it. The off-line input can be carried out quickly and simply. Because of the modular set-up and the transferal of intelligence to the peripheral monitors, the independence of each step of components from the next highest level is achieved [10]. Therefore, if the next step should break down, the patient monitoring remains primarily undisturbed. The analysis of complicated biosignals remains in the sphere of the periphery, so that the microsoftware is simplified [3], and there is much less of a burden on the microprocessor. Here, the transformation of mechanical amplitudes into electrical impulses, and the feeding of these into the microprocessor via the analog-digital transformer, is no problem. Difficulties only arise with the variety of data to be recorded.

Thus, for example, for a sufficiently precise analysis of the course of the respiratory pressure during a respiratory cycle, an intake of 1000 points with a scanning frequency of 100 Hz is required, assuming a pressure curve analysis of the systemic pressure of 1000 Hz.

The subsequent analysis of the curves with an artefact and alarm signal is a heavy burden on the computer which is not to be underestimated. Thus prior preparation through the periphery is a sensible and adequate solution, especially if application of the system is planned in several operating theatres.

The documenting and storing of data in the form of medical records, which is necessary for trouble-free information flow to the subsequent patient treatment, is made possible by the hard-copy unit.

References

1. Apple HP (1980) Automatic noninvasive blood pressure monitors: What is available? In: Gravenstein JS (ed) Essential noninvasive monitoring in anesthesia. Grune Stratton, New York London
2. Bartels H, Adolf J, Bonke St. Maurer PC (1979) Einsatz eines rechnergestützten Überwachungs- und Dokumentationssystems in der postoperativen Behandlung von Risikopatienten. Intensivbehandlung 4:99
3. Bender HJ (1981) Implementierung eines rechnergestützten Patientenüberwachungssystems. Diss. Mannheim
4. Comerchero H, Vernia M, Tivig G, Kalinsky D, Miller A (1979) Solo: An interactive microcomputer-based bedside monitor. 3rd Symposium "Computer applications in medical care". Washington, USA
5. Eberlein HJ (1977) Definition der Invasivität: In: Refresher course. ZAK-Geneve. Medicine et Hygiene, Geneve
6. Ehlers CT (1979) Datenverarbeitung im Klinikum der Georg-August-Universität Göttingen. Beschreibung des Gesamtsystems Göttingen
7. Fournell A, Schwarzhoff W, Steinhoff H, Falke K (1981) Technik der blutigen arteriellen und venösen Druckmessung auf einer Wach- und Intensivstation. In: Epple E, Junger H, Blischer W, Schorer P, Apitz J, Faust U (Hrsg) Rechnergestützte Intensivpflege. Thieme, Stuttgart
8. Gessner U (1979) Fehlerquellen bei der Berechnung von Lungen- und Herzfunktionsstörungen. Biotechn Umsch 3:72
9. Hossli G (1977) Ethische Überlegungen – Coreferat. In: Refresher Course. ZAK Geneve. Medicine et Hygiene, Geneve
10. Kieninger E (1980) Mikroprozessorsystem zur Erfassung und Darstellung von Vitalparametern mechanisch beatmeter Patienten. Diplomarbeit des Studiengangs Medizinische Informatik der Universität Heidelberg, Fachhochschule Heilbronn

11. Klain MM, Finestone SC (1980) Computerized cardio-pulmonary monitoring in the operating rooms. Symposium "Computer in critical care and pulmonary medicine". June 1980. Lund. Sweden
12. Luff NP, White DC (1981) Evaluation of the EMMA anaesthetic gas monitor. Br J Anaesth 53:1102
13. Lustig IJ, Parrish JN, Augenstein JS, Civetta JM, Rodman GA, Caruthers TE (1981) Clinical experimence with a minicomputer based data management system in surgical intensive care. 3rd Intern Symposium "Computers in critical care and pulmonary medicine". June 1981. Norwald, USA
14. Norlander OP (1973) Patientendatensystem for Operation und Intensivpflege. Chirurg 44:446
15. Nunn JF (1969) Applied respiratory physiology with special reference to anaesthesia. Butterworths, London
16. Olsson SC (1980) Clinical studies of gas exchange during ventilatory support – a method using the Siemens-Elema CO_2 Analyser. Br J Anaesth 2:491
17. Opderbecke HW (1981) Der Verantwortungsbereich des Anaesthesisten. In: Opderbecke HW, Weißauer W (Hrsg) Forensische Probleme in der Anaesthesiologie. Perimed, Erlangen
18. Paulsen AW, Frazier WT, Harbort RA, Hartney KJ (1980) Computer aided monitoring for the Anesthetist. Symposium "Computers in critical care and pulmonary medicine". June 1980. Lund, Sweden
19. Pettersson SO, Seeman T, Wahlberg K, William-Olsson G, Ackerhammer E, Öberg PE (1975) The computer in the hospital service, clinically oriented information system. Östra Hospital Gothenburg. Proposal December 1975
20. Rader C, Taylor W, Hansen D (1981) A distributed microprocessor respiratory intensive care monitoring system with mass spectrometer proximal flowmeter and airway pressure transducer. 3rd Intern Symposium "Computers in critical care and pulmonary medicine". June 1981, Norwald USA
21. Ribbe T, Hallen B, Lumarsson D, Nygren G, Norlander O (1980) Data log system for monitoring during anaesthesia. Symposium on "Computers in critical care and pulmonary medicine". June 1980. Lund, Sweden
22. Salat H (1974) Elektronische Patientenüberwachung in der internistischen Intensivpflegestation des Kreiskrankenhauses Herford. Röntgenstrahlen 30
23. Turney SZ (1981) Computerized multibed respiratory monitoring. 3rd Intern Symposium "Computers in critical care and pulmonary medicine". Norwald, USA
24. Zeelenberg C, Hoare MR (1981) Herzrhythmusüberwachung. In: Rechnergestützte Intensivpflege. INA. Bd 26. Thieme, Stuttgart New York

Discussion XVI

Faithfull: You mentioned the possibility of off-line input into the system. Does this enable you to receive calculated values, i.e. cardiac input, cardiac output, left ventricular-work?

Weller: We can measure only oxygen capacity, vessel resistance and pulmonary pressure with a Swan-Ganz catheter.

On-Line Graphic Presentation of Lung Mechanics During Mechanical Ventilation

O. Prakash, B. van der Borden, and S. Meij

The airway of the patient on mechanical ventilation is an ideal place for safe, non-invasive measurements of airway pressure, flow, CO_2 concentration and oxygen uptake. With advanced technology, these measurements can be automated and presented to the physicians in numerous forms.

The information about alveolar ventilation, compliance, resistance, end-tidal CO_2, tidal CO_2 production, pressure/volume, pressure/flow and volume/CO_2 concentration and CO_2 production graphic plots are useful for assessment of the physiological condition of the lung on a minute-to-minute basis.

The most commonly used presentation is done in the form of numerical values on digital displays or as readings on analogue meters. As an extension, a few of the monitored parameters can be recorded as function of time on a paper recorder confirming the reliability of the numerical values. The disadvantage of the system is the collection of huge piles of recorder paper for manual interpretation. A good alternative would be to display different parameters, e.g. airway pressure and carbon dioxide concentration, against time on a display unit on a continuous basis. If a display unit is used, one has the possibility of producing graphic plots in the form of half-loops, such as pressure/volume or a single-breath carbon dioxide loop, which may be useful in early detection on difficult situations during respiratory care (Fletcher; Osborn 1977). To investigate the usefulness of graphic presentation, a system is developed to present real time half-loops of inspired airway pressure (cm H_2O) against inspired volume; expired airway flow (l/s) against ex-

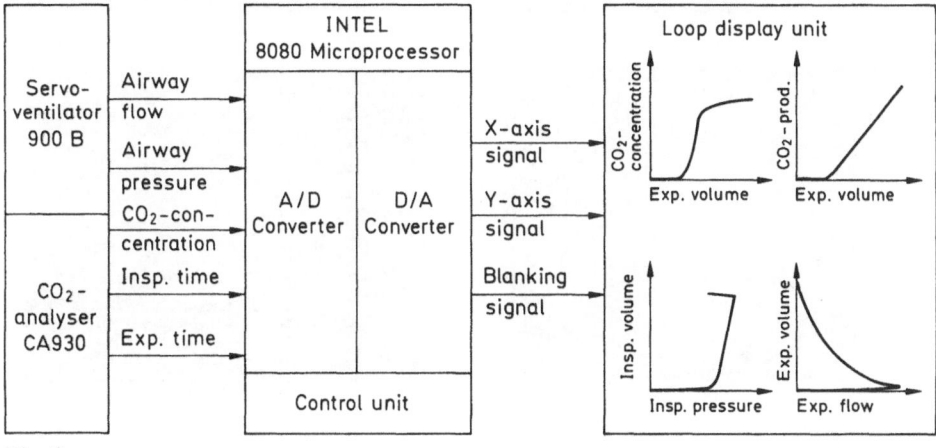

Fig. 1

225

pired volume, CO_2 concentration (%) against expired volume and CO_2 production (ml) against expired volume on an x-y oscilloscope.

A microprocessor (INTEL 8080) is used together with a four-channel analogue to digital converter, a three-channel digital to analogue converter and an x-y oscilloscope. The analogue to digital converter forms the interface between the ventilator (S.E. 900B) and carbon dioxide analyser (CA930) and the microcomputer. The digital to analogue converter forms the interface between the microcomputer and the x-y oscilloscope (Fig. 1). The output signals from the ventilator and carbon dioxide analyser, such as airway pressure, airway flow, CO_2 concentration, CO_2 production and timing signals, are fed into the microprocessor system and sampled at a sampling rate of 40 Hz. Using the above-mentioned signals it is possible to produce a breath-by-breath display of four half-loops.

On-line, a half-loop of one breath is built up on the screen and remains visible until a new breath starts. The still visible half-loop is then overwritten by the new half-loop on the screen. It is possible to freeze a half-loop on the screen and photograph it with a polaroid camera. Only one loop can be displayed on the screen at a time. A selection of four different half-loops is made by a switch on the front panel. The loops are respectively inspired airway pressure (cm H_2O) against inspired volume and expired airway flow (l/s) against expired volume, giving information regarding compliance and resistance (Osborn 1977), and CO_2 concentration (%) against expired volume (ml) and CO_2 production (ml) against (Figs. 2–4) expired volume (ml), giving information about gas exchange and dead

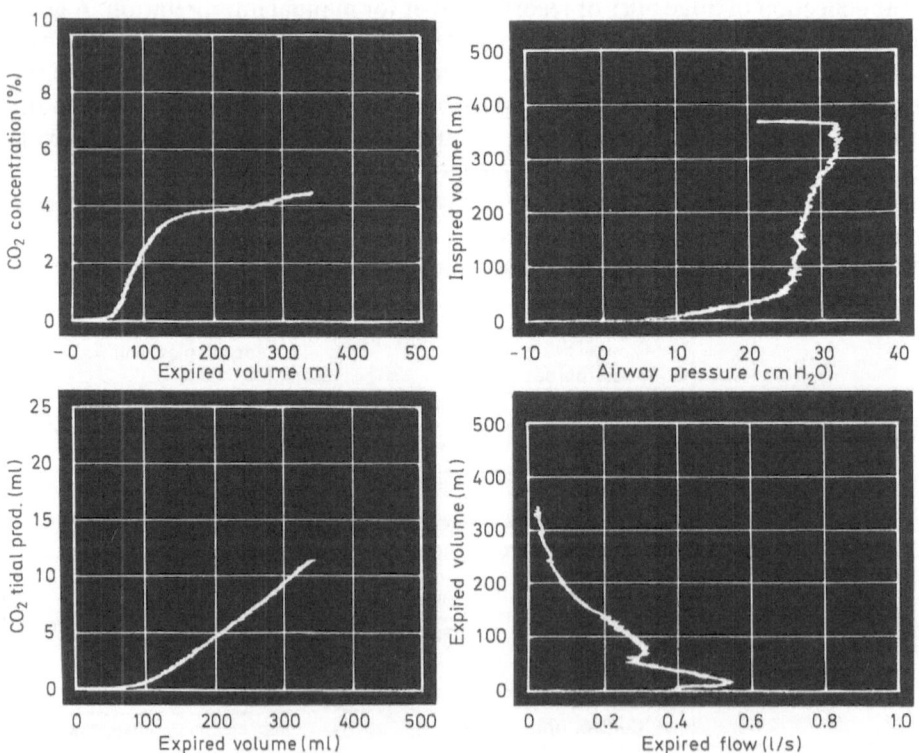

Fig. 2. Illustration of normal pressure/volume and flow/volume curves

226

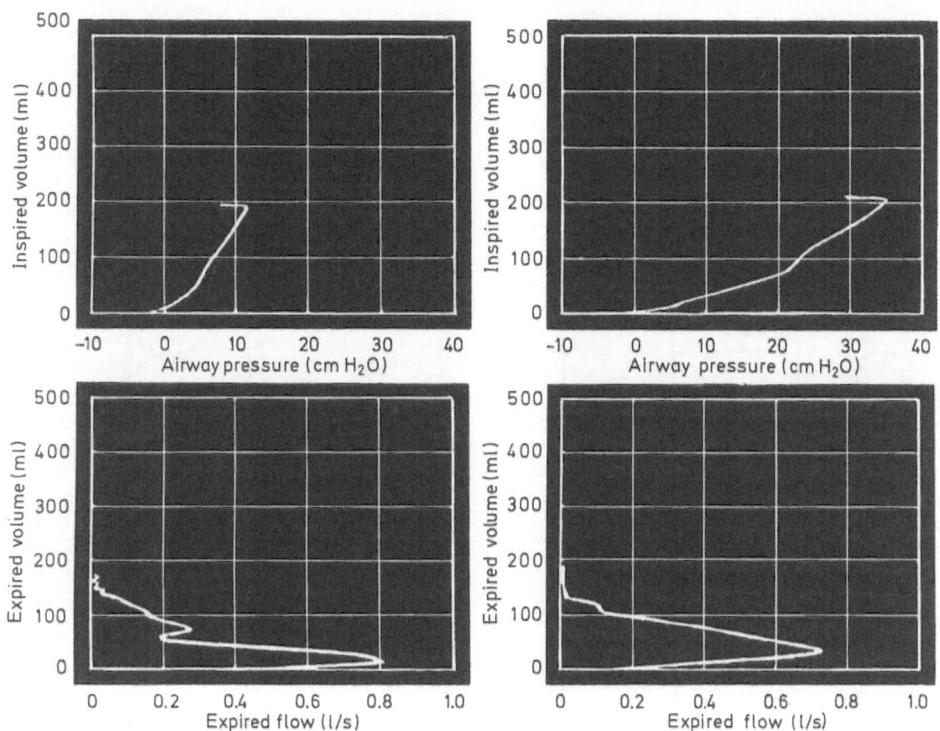

Fig. 3. A patient in whom high airway resistance developed due to pulmonary oedema. The pressure/volume curve shows an over-extended belly to the right and flow/volume plot shows lower than normal peak flow

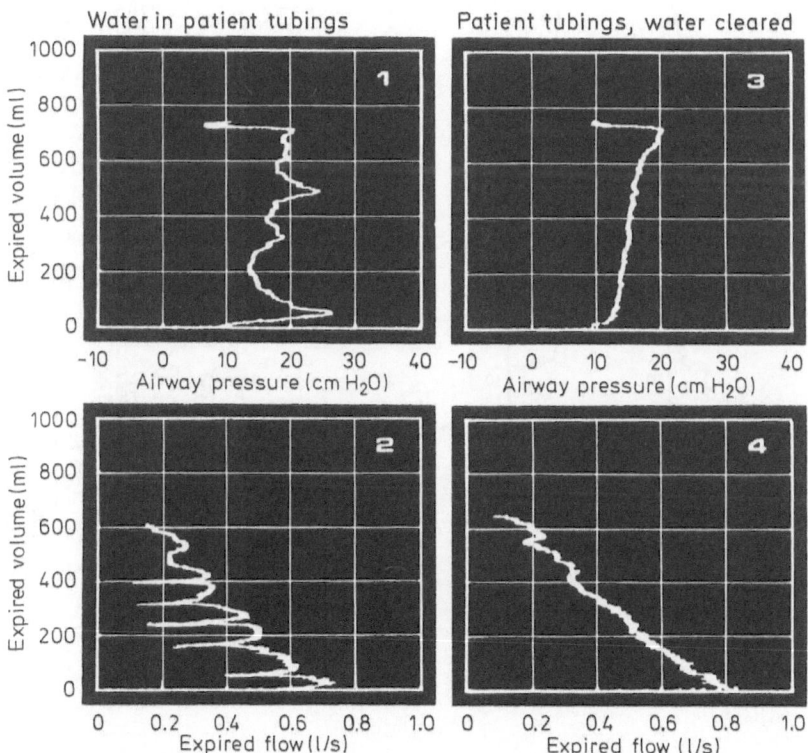

Fig. 4. An example of fluid in the airway or ventilator hoses

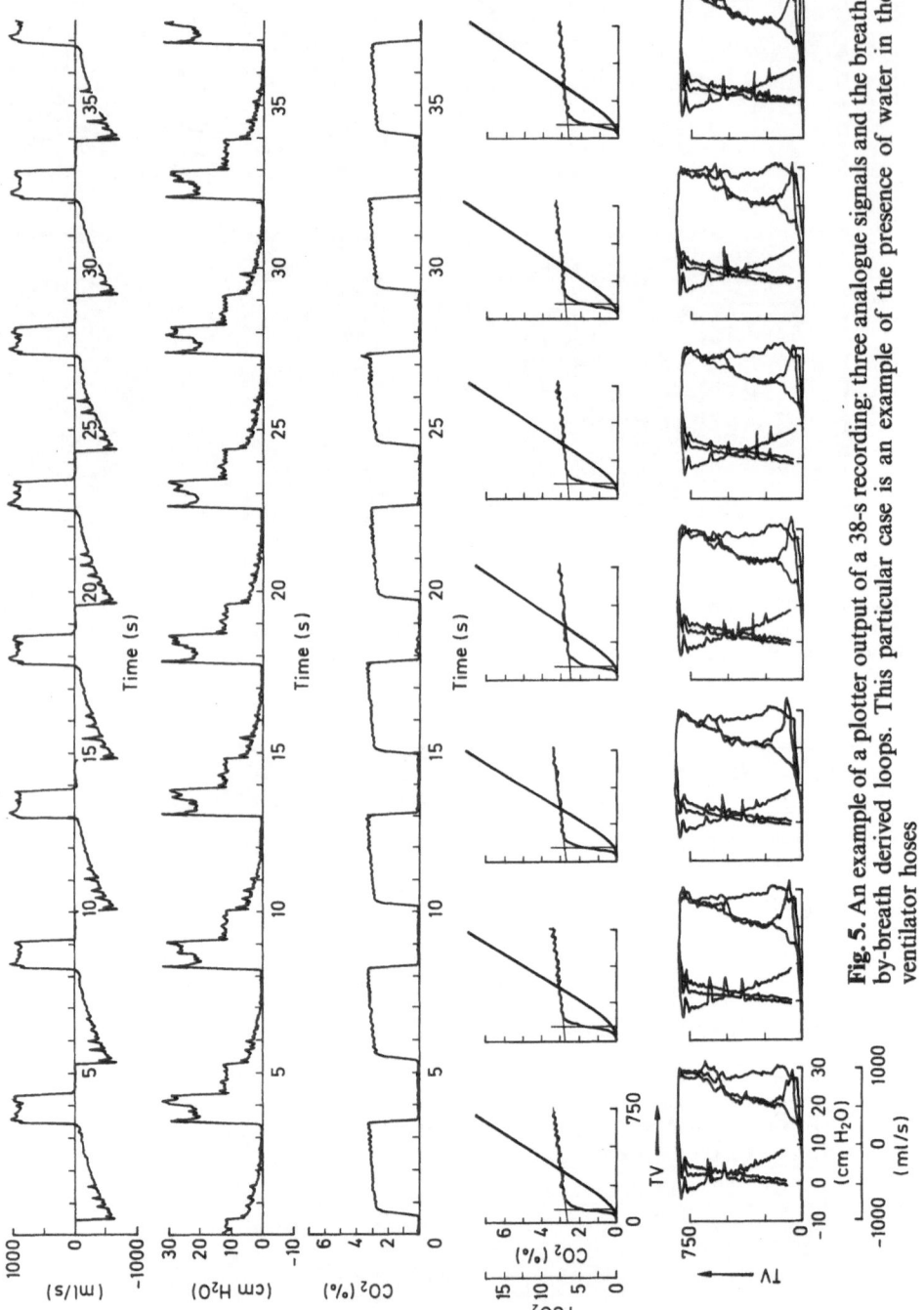

Fig. 5. An example of a plotter output of a 38-s recording: three analogue signals and the breath-by-breath derived loops. This particular case is an example of the presence of water in the ventilator hoses

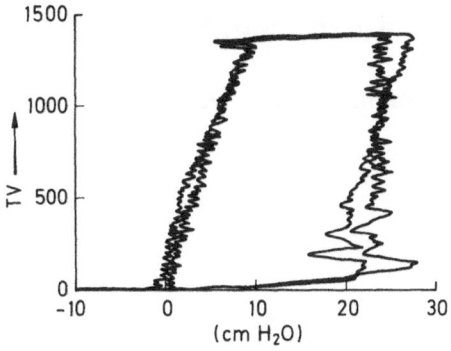

Fig. 6. An example of a pressure/volume loop. The illustration demonstrates a measured loop compared to a calculated loop

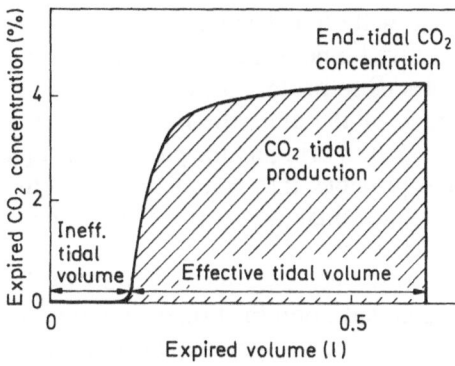

Fig. 7. Single-breath CO_2 curve. Expired CO_2 concentration versus expired volume (litres); *shaded area* shows the effective tidal volume ventilation and tidal CO_2 production. The initial carbon-dioxide-free part of the tidal volume represents gas compressed in the ventilator tubing and humidifier and gas from the conducting airways, i.e. ineffective tidal volume

space (Fletcher 1979). An automatic scaling of the displayed loops had yet to be implemented. At present the selection of the correct scale is done by turning a knob on the front panel of the microprocessor system.

Additionally, all the analogue signals mentioned above (airway flow, airway pressure and CO_2 concentration), together with the timing signals of the ventilator, are recorded on an eight-channel analogue tape-recorder (HP 396BA). The signals are then sampled on a larger system (PDP 11/34). This recording system is mainly used for research purposes and for evaluation of the microprocessor system described above. The signals can be plotted on a four-colour plotter (HP 7221B) together with the single-breath CO_2 curves (CO_2% against expired tidal volume and tidal CO_2 production against expired tidal volume), pressure/volume and flow/volume loops (Fig. 5). Two pressure-volume loops are plotted: one from the measured pressure and flow signal and one from the calculated pressure and measured flow signal (Fig. 6). Comparison of the two loops gives information on the reliability of the computed resistance and compliance from the linear model: $P = V/C + \dot{V}.R.$ (Osborn 1977). The two different methods of obtaining airway dead space are indicated in the CO_2 single-breath curve (Fig. 7) (Fletcher 1979).

References

Fletcher R (1979) The single breath test for carbon dioxide. Thesis, Lund, Sweden
Osborn JJ (1977) Cardiopulmonary monitoring in the respiratory intensive care unit. Med Instrum 11:278–282

Check-List for the Dräger Narkosespiromat 650

V. Rejger, A. v. Bijnen, D. de Haas, H. J. Beekman, and F. de Raadt

Only properly functioning modern equipment will offer the anaesthesiologist security during his daily work. By providing active homeostatic monitoring during anaesthesia and surgical procedures it enables us to protect the patient from the effects of surgical trauma. Although the task of the anaesthesiologist and the variety of equipment available are meant to assist in bringing the patient through the operative phase in as good a condition as possible, they can at the same time form a threat to him (electrocution, burns, explosions). Apart from "human errors" technical malfunctions are beginning to play an increasingly important role in patient safety. We are of the opinion that at the present time the safety of the patient in the operating-theatre is in part dependent upon the proper functioning of the equipment serviced by expert technicians, and this plus adequate apparatus selection will result in a decrease in the frequency of human error.

In the past, attempts made by anaesthesiologists or their nurses to solve technical problems arising in anaesthetic machines have not always ended successfully. More often than not, their inept manipulations have resulted not only in danger to personnel and patient, but also in damage to the apparatus itself. Frequently, the anaesthetic machines became so defective that it was necessary to remove them from active duty for long periods of factory repair. For this reason alone, it was necessary for hospitals to maintain large reserves of anaesthetic machines. In addition, biannual servicing offers no protection if these machines are used intensively every day, and the service costs are certainly not minimal. The small defects and damaged parts which cannot be repaired or replaced without factory servicing have been largely responsible for compromising the investments made by hospitals in modern anaesthetic machines, since a number of them cannot be used daily. Because of these situations and due to the understanding nature of our hospital's medical, economic and technical departments we have been able to create a new organization for the control, maintenance and care of our anaesthetic machines.

Since our hospital is organized on a pavilion system, the surgical blocks, operating theatres, recovery rooms and treatment rooms are spread throughout the whole hospital terrain. Consequently, it was necessary to create a central technical maintenance department which would solely occupy itself with the maintenance and repair of monitors, ventilators and anaesthetic machines. This department is open 24 h a day.

In one of the surgical blocks, which contains 12 operating theatres and one recovery room, we have located an emergency technical maintenance room. This area is continually manned by technicians throughout the whole of the daily surgical programme. These people perform minor, short-lasting anaesthetic machine repairs in the operating theatres during the surgical procedures. If more extensive repairs are necessary they then replace the defective machines with reserve apparatus and transfer them to the technical maintenance area for immediate repair. Since a large portion of the repairs and maintenance work are performed by this well-equipped service it is rarely necessary to send anaesthetic machines on to the factory.

Due to the excellent collaboration between the medical and technical staffs, the technical shortcomings of the anaesthesiologists and nurses are readily compensated by the knowledge and experience of the technicians. In turn, the technicians have gained an appreciation of the medical aspects related to their work and this is reinforced by the daily contact between the two groups. In addition, as a direct result of these technical controls and repairs the incorrect usage of the anaesthetic machines has been greatly reduced. In the course of years it became evident to us that not only the technicians but also the anaesthetic team should be responsible for the proper functioning of equipment. In view of the increasing technical knowledge being acquired by the anaesthesiologists there appeared to be no reason why they should not be able to evaluate the proper functioning of their anaesthetic machines and even, if necessary, correct simple malfunctions. Consequently, our anaesthesiologists and nurses regularly receive technical information and instruction.

In order to technically evaluate the functional state of an anaesthetic machine, a check control procedure developed by the manufacturer is necessary. These procedures are very extensive and must always be performed by specially trained technicians. There is little difference between the importance of the anaesthetic machine, monitors and other equipment found in the operating theatre and that of the equipment in the cockpit of an aeroplane. We can only state that our machines are technically less safe and still not computerized (fully automatic controls). One may ask, is the life of a patient in the operating theatre less valuable than that of 300 passengers?

At present there are 26 physicians and 12 nurses in training in our Anaesthesiology Department. In order to minimize the number of human and technical errors it is essential that our anaesthetic machines be meticulously and routinely checked. Consequently, we have directed our efforts towards achieving the following goals:

1. All anaesthetic machines must be checked both before and at the end of the daily operating-programme.
2. The complex manufacturer's check-list should be supplemented by a simpler procedure which guarantees the daily safety of the anaesthetic machine.
3. The check-list procedure should take no longer than 10 min to perform.

Due to good collaboration between Dräger, Holland, and our technical anaesthetic team, a simple check-list which enables controls to be performed without the need of complex instruments was developed within a short period of time. The use of this check-list is demonstrated in Fig. 1. This check-list consists of ten points (see Table 1) which, if observed in order, guarantee that the Narkosespiromat 650 will function properly. This is essential because should unexpected patient reactions manifest themselves in the course of anaesthesia we must be certain of the reliability of our anaesthetic machine.

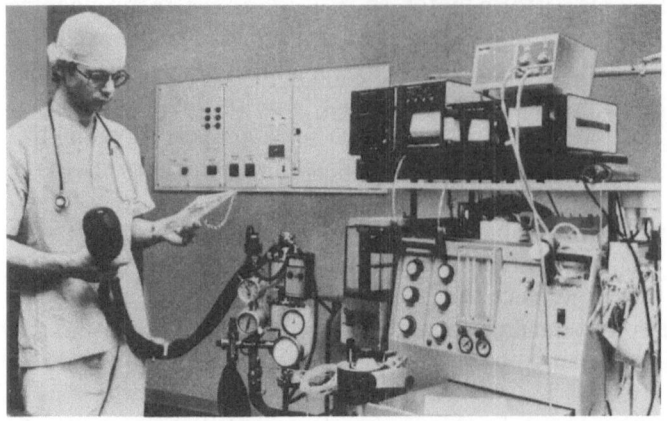

Fig. 1. Anaesthesist performing leakage test according to the checklist

Table 1. Control procedure for the Dräger Narkosespiromat 650

---

1. Pressure control
   Remove O_2 and N_2O lines from central supply unit. Ventilation half-open.
   Machine on automatic   Inspiration 60 Mbar.   Expiration O.
   Volume maximal.   Insp.: exp. = 1:2.   Trigger O.
   Respiratory pressure gauge O. Overflow valve closed. Start machine. Resp. pressure
   gauge must demonstrate a pressure > 50 Mbar within 25 s.

2. Anaesthetic gas supply
   Check:
   a) O_2 (oxygen)
      O_2 line – connect to central supply unit.
      O_2 rotameter – open to 2 l/min.   Bobbin must rotate.
      N_2O (nitrous oxide) – open rotameter.   Bobbin must not rise.
      Oxygen concentration monitor – must register > 60% O_2
   b) $O_2 - N_2O$ safety control
      N_2O line – connect to central supply unit.
      N_2O rotameter – open to 4 l/min.   Bobbin must rotate.
      O_2 line – disconnect. N_2O bobbin must fall (before O_2).
      Alarm must activate.

3. Bypass
   Open bypass. O_2 flow must be audible.
   $O_2 - N_2O$ safety device must not be activated.
   Oxygen concentration monitor must register > 60% O_2. Close bypass.

4. Vaporizers
   Check:
      Filling level.
      Selection switch (arrow upwards during rotation).

5. Vacuum
   Open valve. Close off tube. A vacuum must arise.

6. Circle system
   Check:
   a) Tubes, connections, valves, valve heads
   b) Absorbers
   c) Volumeter (zero calibration, clock, bypass)
   d) Respiratory pressure gauge (zero calibration, alarm, alarm threshold)
   e) Blood pressure meter (zero calibration, function; leak of 10 mmHg permitted)
   f) Clock (wind-up, adjust time, stop-watch function).

Table 1 (contd.)

___

7. Reservoir balloon
   Check:
   Balloon (leakage), fitting (screwed in properly).

8. Function control
   Check:
   a) Control head (balloon, bellows, must not touch plastic hood)
   b) Inspiratory pressure

Volume maximal.	Expiration O.	Trigger O.
Machine automatic.	Frequency 12.	Ventilation half-open.
E- or Y-piece – close off.	Start machine.	

   Inspiratory pressure: Minimal registered pressure 5–15 Mbar; 40 Mbar: 38–44 Mbar; 60 Mbar: 58–65 Mbar.

   c) Expiration

Volume 500.	Trigger 0.	Insp.: Exp. = 1:2.

   Machine automatic. Frequency 12. Ventilation half-open.
   Connect training thorax. Start machine.
   Expiration maximally positive: Registered pressure +10 to +22 Mbar.
   Expiration maximally negative: Registered pressure −10 to −20 Mbar.

   d) Frequency
   Adjust machine as before. Inspiration 40 Mbar.
   Frequency maximal. Bellows must rise and fall.
   Frequency minimal. Bellows must rise and fall.
   Frequency 30. Bellows must give 28–32 cycles/min.

   e) Inspiration: Expiration.
   Adjust machine as before. Frequency 12.
   Inspiration: expiration = 1:2. Control with a stop-watch.

   f) Trigger
   Adjust machine as before. Insp.: exp. = 1:4.
   Trigger maximal. Sensitivity maximal without self-triggering.
   Trigger minimal. Machine does not trigger.

9. Leakage test
   a) Automatic:

Volume 500. Inspiration 40 Mbar.	Expiration maximal.
Ventilation half-closed.	Frequency minimal.
Insp.: exp. = 1:2. Trigger O.	Machine on.

   O_2 rotameter opened, until resp. pressure meter +15 Mbar.
   O_2 flow must be 0.8 l/min.

   b) Manual ventilation:
   Adjust machine as before. Machine to manual ventilation.
   O_2 rotameter opened until resp. pressure meter +40 Mbar.
   O_2 flow must be 1.2 l/min.

10. Delivery state
    Connect electrically. O_2 and N_2O connected to central supply unit. Cylinders closed. Connect fresh gas line. Overflow valve circle system closed. Ventilation half-closed. Machine on automatic. Volume 500. Inspiration 40 Mbar. Expiration 0. Insp.: exp. = 1:2. Frequency 12. Trigger 0. On/off switch off.

___

Our checking procedure is as follows:

1. Every morning, before the start of the surgical programme, the anaesthesiologist checks his anaesthetic machine.
2. At the end of the day's surgical programme it is the anaesthetic nurse who checks the machine (double-check).
3. All the performed checks are confirmed in writing.

Table 2. Check-list for the Narkosespiromat 650

Time					Remarks
Pressure control					
Anaesthetic gases suppl.					
Oxygen safety device					
Bypass					
Vaporizers					
Vacuum					
Tubes, valves, connection					
Absorbers					
Volumeter					
Resp. pressure gauge					
Blood pressure meter					
Clock and stopwatch					
Reservoir balloon					
Control head					
Inspiration pressure					
Expiration pressure					
Frequency					
Insp.: exp.					
Trigger					
Leakage test automat					
Leakage test ventilation					
Delivery state					
Checked by:					

Series no. _____ Date _____

4. All abnormalities are noted on the check-list (see Table 2) and are immediately reported to the technical maintenance department.

A defective anaesthetic machine may not be used until all faults are corrected. If extensive repairs are necessary it will be replaced by a reserve apparatus. In addition, the technical maintenance department regularly carries out "spot controls" to determine whether the checking procedures are being performed properly and if the anaesthetic machines are truly in good working order.

Other aspects arising out of this new procedure are:

1. Due to the introduction of the check-list it is no longer necessary to have the anaesthetic machines checked out in the Anaesthesia Service Centre. Instead, after disinfection the anaesthetic machine is immediately checked out by either the anaesthesiologist or anaesthetic nurse upon its return to the operating-theatre. This has the advantage of eliminating possible defects which may arise in the time elapsing between its control in the Anaesthesia Service Centre and its use in the operating theatre.

2. Written documentation is of inestimable value upon the occurrence of medical accidents. Under the circumstances of a medico-legal process the check-list provides an accurate, objective piece of evidence.

After working with the check-list for 2 years we are of the opinion that:

1. The anaesthetic machine has become safer and thereby less of a threat to the patients.
2. By its use anaesthesiologists have gained more insight into the functioning of the anaesthetic machine, so that they are now able to solve minor malfunctions easily and rapidly without the assistance of technicians.
3. The investment in reserve apparatuses by the hospital can be decreased.
4. Technical repairs are no longer the cause of interruption or delay in the operating programme.
5. The check-list has decreased the workload of the Anaesthesia Service Centre.
6. In the case of accidents there is accurate documentation available about the functioning of the machine used by the anaesthesiologist.
7. Our future task will be to include all the anaesthetic equipment at the disposal of the anaesthesiologist in a check-list control procedure.

Closing Statement

R. Droh

Ladies and gentlemen, I would like to thank all of you: the interpreters, all authors, the audience, my staff, and all my friends for the excellent co-operation.

We will meet again at the Second International Symposium "Innovations in Management and Technology and Pharmacology" next year.